Springer Theses

Recognizing Outstanding Ph.D. Research

Aims and Scope

The series "Springer Theses" brings together a selection of the very best Ph.D. theses from around the world and across the physical sciences. Nominated and endorsed by two recognized specialists, each published volume has been selected for its scientific excellence and the high impact of its contents for the pertinent field of research. For greater accessibility to non-specialists, the published versions include an extended introduction, as well as a foreword by the student's supervisor explaining the special relevance of the work for the field. As a whole, the series will provide a valuable resource both for newcomers to the research fields described, and for other scientists seeking detailed background information on special questions. Finally, it provides an accredited documentation of the valuable contributions made by today's younger generation of scientists.

Theses are accepted into the series by invited nomination only and must fulfill all of the following criteria

- They must be written in good English.
- The topic should fall within the confines of Chemistry, Physics, Earth Sciences, Engineering and related interdisciplinary fields such as Materials, Nanoscience, Chemical Engineering, Complex Systems and Biophysics.
- The work reported in the thesis must represent a significant scientific advance.
- If the thesis includes previously published material, permission to reproduce this must be gained from the respective copyright holder.
- They must have been examined and passed during the 12 months prior to nomination.
- Each thesis should include a foreword by the supervisor outlining the significance of its content.
- The theses should have a clearly defined structure including an introduction accessible to scientists not expert in that particular field.

More information about this series at http://www.springer.com/series/8790

Xiaoming Chen

Analysis and Synthesis of Positive Systems Under ℓ_1 and L_1 Performance

Doctoral Thesis accepted by
The University of Hong Kong, Hong Kong

Springer

Author
Dr. Xiaoming Chen
College of Automation Engineering
Nanjing University of Aeronautics and Astronautics
Nanjing
China

Supervisor
Prof. James Lam
Department of Mechanical Engineering
The University of Hong Kong
Hong Kong
Hong Kong

ISSN 2190-5053 ISSN 2190-5061 (electronic)
Springer Theses
ISBN 978-981-10-2226-5 ISBN 978-981-10-2227-2 (eBook)
DOI 10.1007/978-981-10-2227-2

Library of Congress Control Number: 2016946620

Printed on acid-free paper

This Springer imprint is published by Springer Nature
The registered company is Springer Science+Business Media Singapore Pte Ltd.

Supervisor's Foreword

In systems and control areas, study on positive systems has drawn much attention from many researchers in recent years. The variables of positive systems involve quantities which are naturally nonnegative. In other words, such systems take nonnegative values at all times since they usually denote the concentrations or amounts of material in application fields. Physical examples of positive systems are abundant in various fields ranging from economy and sociology, to ecology and biology. For instance, the amount of fluids in tanks, the number of species in an ecosystem, and the concentration of substances in chemical processes are all described by nonnegative quantities. Hence, it is of practical meaning to investigate positive systems.

Positive systems are defined on cones rather than linear spaces and possess many unique features. Therefore, many new problems appear and some previous approach used for general systems are no longer applicable to positive systems. For example, in general linear system theory, if a system is controllable the poles of the system can be placed arbitrarily, whereas for positive linear systems, this feature may not be true owing to the positivity constraints on the system matrices. In view of the widespread applications and special characteristics of positive systems, it is necessary to investigate the analysis and synthesis problems for positive systems.

This book introduces novel and significant results regarding the analysis and synthesis problems for positive systems under ℓ_1 and L_1 performance. Two classes of systems are considered. One is positive linear system and the other is positive fuzzy system. It describes stability analysis, controller synthesis, and bounding positivity-preserving observer and filtering design for a variety of both discrete and continuous positive systems. Some widely encountering factors such as time-delay and uncertainties are taken into consideration. In this book, the computationally efficient solutions are derived based on linear programming in terms of matrix inequalities, and a number of analytical solutions are obtained for special cases. In all, this book applies a series of novel approaches and fundamental techniques to the further study of positive systems, thus contributing significantly to the theory of positive systems which is a hot area in the control field.

Hong Kong
June 2016

Prof. James Lam

Parts of this thesis have been published in the following articles:

1. Chen X, Lam J, Li P et al (2013) ℓ_1-induced norm and controller synthesis of positive systems-induced norm and controller synthesis of positive systems. Automatica 49:1377–1385 (Reproduced with Permission).
2. Chen X, Lam J, Li P et al (2014) Output-feedback control for continuous-time interval positive systems under L_1 performance. Asian Journal of Control 16(6): 1592-1601 (Reproduced with Permission).
3. Chen X, Lam J, Li P (2014) Positive filtering for continuous-time positive systems under L_1 performance. International Journal of Control 87(9): 1906-1913 (Reproduced with Permission).
4. Chen X, Lam J (2014) Positive state-bounding observer for interval positive systems under L_1 performance. In Proceedings of the 33rd Chinese Control Conference (Reproduced with Permission).
5. Chen X, Lam J, Lam HK (2015) Positive filtering for positive Takagi-Sugeno fuzzy systems under ℓ_1 performance. Information Sciences 299: 32-41 (Reproduced with Permission).

Parts of this thesis have been published in the following articles:

1. [illegible]

2. [illegible]

3. [illegible]

4. [illegible]

Acknowledgments

I would like to express my sincere thanks to my supervisor, Prof. James Lam, for his enlightening guidance, insightful ideas, and endless support during my candidacy at the University of Hong Kong. His rigorous research ethics, diligent work attitude, and wholehearted dedication to his students have been an inspiration to me and will influence me forever.

My sincere thanks also go to my former supervisor, Prof. Huijun Gao for his guidance and support. The discussion with him aroused my interest in control theory and broadened my horizons. He always encourages me and supports me even after I graduated from the Harbin Institute of Technology, for which I am truly grateful. I am grateful to Dr. Hak-Keung Lam at King's College London for inviting me to visit his group. I would like to thank the students and research staffs in the control laboratory. The friendships with them make my student life at HKU very colorful.

Finally, I wish to express my deepest gratitude to my parents, my younger sister and my husband for their love, encouragement, and support. I sincerely dedicate this thesis to them.

The writing of the book is financially supported in part by the National Natural Science Foundation of China under Grants 61503184, 61573184, and 61503037, the Scientific Research Funds of Nanjing University of Aeronautics and Astronautics 56XCA15013.

Acknowledgments

Contents

Acronyms

BMI	Bilinear Matrix Inequality
ILMI	Iterative Linear Matrix Inequality
LMI	Linear Matrix Inequality
MIMO	Multi-Input Multi-Output
SIMO	Single-Input Multi-Output
SOF	Static Output-Feedback
T–S	Takagi–Sugeno

Symbols

$\mathbb{R}$	Set of real numbers
$\mathbb{R}^n$	Set of n-column real vectors
$\bar{\mathbb{R}}^n_+$	Set of n-dimensional nonnegative real vectors
$\mathbb{R}^{n\times m}$	Set of $n \times m$ real matrices
$\\|x\\|$	Euclidean norm of the vector x
$\\|x(k)\\|_1$	$\sum\limits_{i=1}^{n} \|x_i(k)\|$
$\\|x(t)\\|_1$	$\sum\limits_{i=1}^{n} \|x_i(t)\|$
$\\|x\\|_{\ell_1}$	$\sum_{k=0}^{\infty} \\|x(k)\\|_1$
$\\|x\\|_{L_1}$	$\int_0^{\infty} \\|x(t)\\|_1 dt$
$\\|Q\\|_1$	$\max\limits_{1\leq j\leq n} (\sum\limits_{i=1}^{m} \|q_{ij}\|)$
$\\|\mathcal{G}\\|_{(\ell_1,\ell_1)}$	ℓ_1-induced norm of the operator $\mathcal{G}$
$\\|\mathcal{G}\\|_{(L_1,L_1)}$	L_1-induced norm of the operator $\mathcal{G}$
ℓ_1	space of all vector-valued functions with finite ℓ_1 norm
L_1	Space of all vector-valued functions with finite L_1 norm
$\in$	Belong to
$\triangleq$	Defined as
$\square$	End of proof
I	Identity matrix
A^T	Transpose of the matrix A
A^{-1}	Inverse of the matrix A
$\mathrm{diag}(A_1, \ldots, A_n)$	Block diagonal matrix with $A_1, \ldots, A_n$ on the diagonal
$\mathrm{col}_i(A)$	The *ith* column of matrix A
$\rho(A)$	$\max\{\|\lambda_i(A)\|, i = 1, 2, \ldots, n\}$, i.e., spectral radius of matrix A
$\alpha(A)$	$\max\{\mathrm{Re}\lambda_i(A), i = 1, 2, \ldots, n\}$, i.e., spectral abscissa of matrix A
$\\|A\\|$	Spectral norm of the matrix A
$A > B$	$A - B$ is positive definite

$A \geq B$	$A - B$ is positive semi-definite
$A \geq\geq B$	$A - B$ is element-wise nonnegative
$A >> B$	$A - B$ is element-wise positive

List of Figures

Chapter 1
Introduction

1.1 Background

A control system is an interconnection of components forming a system configuration that will provide a desired system response. It is a arrangement of physical components and is composed of inputs, outputs and state variables. An input is a channel where changes can be injected into a system so as to activate or manipulate a process. An output is a channel where the response can be measured or observed. A state is a set of mathematical functions or physical variables which can be used to describe completely the future behaviour of a dynamic system if all the inputs future inputs are known. In many practical systems, there is such a kind of systems whose state variables and outputs are always positive (or at least nonnegative) for nonnegative inputs and initial conditions. These systems are commonly referred to as positive systems.

Filtering is one of the basic problems in the fields of systems, control and signal processing. The goal of filtering is to estimate signals that are unmeasurable through processing the measured output signals. Since the development of Kalman's optimal filtering theory for stochastic systems and Luenberger's observer theory for deterministic systems, many scholars have devoted considerable efforts to the problem of filtering. Especially, the well-known Kalman filtering theory plays a significant role in various social and engineering fields such as aerospace, astronautics, industrial processes and economic and financial systems. The Kalman filtering theory is based on the availability of the precisely-known mathematical model of the studied plant and the assumption of strict Gaussian random processes or series. However, it is usually difficult to characterize the dynamics of the studied plant exactly by a mathematical model, inevitably leading to an error between the derived mathematical model and the practical plant; moreover, it is rare for practical external noises to completely satisfy the strong Gaussian assumption. The uncertainties existing in systems and signals would greatly degrade the performance of a traditional Kalman filter and even cause divergence. Hence, it is of practical meaning to research filtering

X. Chen, *Analysis and Synthesis of Positive Systems Under ℓ_1 and L_1 Performance*, Springer Theses, DOI 10.1007/978-981-10-2227-2_1

theory for uncertain dynamical systems so as to improve the robustness of a filter against uncertainties.

Physical examples of positive systems are abundant in various fields ranging from economy and sociology, to ecology and biology. For instance, the amount of fluids in tanks, the number of species in an ecosystem, or the concentration of substances in chemical processes are all described by non-negative quantities. However, a lot of well-established results for general linear systems cannot be readily applied to positive systems due to the fact that positive systems are defined on cones rather than linear spaces. In fact, in general linear system theory, if a system is controllable, the poles of the system can be placed arbitrarily, whereas for positive linear systems, this feature may not be true owing to the positivity constraints on the system matrices.

In view of the widespread applications and special characteristics of positive systems, it is necessary to investigate the analysis and synthesis problems for positive systems. Like general systems, the goal of control science and engineering for positive systems is to determine the control input of a positive dynamic system such that the system output behaves in a specific desirable way. For example, we may want the temperature and pressure in a reactor vessel in a chemical process plant to be maintained at desired levels. Stability, disturbance attenuation and robustness are three essential issues that should be taken into account thoroughly.

In the control-theoretic context, stability is always given the top priority. Stability means that the steady state of a system is insensitive to the small perturbations of initial conditions, which is usually termed as Lyapunov stability. Roughly speaking, a system is Lyapunov stable if any trajectory starting close enough to the equilibrium will remain close forever. Since the main limitation of Lyapunov stability is that the system trajectory is not required to converge to zero, it may not be enough for practical applications. A stronger concept called asymptotic stability is proposed. A system is said to be asymptotically stable if it is stable and furthermore its trajectory converges to the equilibrium as time approaches to infinity. The main limitation of this concept is that the transient process is not described in a precise way and the transient property is often significant in engineering applications. Therefore, a refined concept, called as exponential stability is offered. It indicates that the convergence is not slower than a particular exponential envelop. In addition to Lyapunov stability, other stability concepts such as bounded-input bounded-output stability are also significant in engineering applications.

The performance of a dynamic system is usually characterized by the attenuation ability against the disturbance inputs, and frequently quantified by some norms related to the system responses and the disturbance inputs. A popular tool to quantify the system performance is the H_∞ norm, which can be termed as the L_2-induced norm as well. The H_∞ norm represents the maximum gain of the system which characterizes the worst-case norm of the regulated outputs over all exogenous inputs with bounded energy. Another commonly adopted measure is the H_2 norm, which quantifies the output variance of the system with white noises as the exogenous input. For positive systems, ℓ_1 and L_1-induced norm can provide a more natural description because 1-norm gives the sum of the values of the components, which is more appropriate, for instance, if the values represent the amount of material or the number of animals in a species.

When modeling physical systems, uncertainties are always inevitable due to some unavoidable factors, such as errors in measurements, limitation in data acquisition and variation of plants. Therefore, it is necessary to consider the effects on stability and performances of the differences between the physical system and the model used. Robustness is another essential specification of control systems and it refers to checking whether the system performance insensitive to the modeling uncertainties. To conduct such an analysis, an uncertain model is often constructed, which generally consists of the real system and uncertainties. Generally, there are two frameworks to characterize the uncertainties. One is the integral quadratic constraints (IQC) description, which is also referred to as the dissipative description, and the other is the structured description, which often includes polytopic uncertainty description and interval uncertainty description. The main task of robust control is to design a controller such that the closed-loop model achieves the required performance and the performance maintains over the admissible uncertainty set.

In the synthesis of dynamic systems, it is often carried out under the state-space framework, and built based on the analysis results. Generally speaking, some performance based criteria should be first obtained under which the controlled/filtered system has the desired properties. Then, a controller or a filter will be designed such that the closed-loop system or the filtering error system satisfies the criterion. Solutions to both the analysis and the synthesis problems are often involved with the solvability of certain matrix equations and/or inequalities.

To achieve the desirable control specifications, the predominant approach is the use of feedback compensation. There are many control strategies such as state-feedback, output-feedback, static control, dynamic control, robust control and adaptive control. Different appropriate control strategies are used according to practical situations. Among these strategies, state-feedback control is usually employed because the state of a system can describe the dynamic behavior completely. In many cases, due to the inaccessibility of the system state, an alternative and more realistic method is static output-feedback (SOF) control. It should be mentioned that many other synthesis problems such as model reduction, filter design can be reformulated as a SOF control problem. For positive system, the positivity of the closed-loop system or the filter is often required, making the synthesis problems for positive systems more complicated than general systems.

1.2 Research Problems

This thesis is concerned with the analysis and synthesis problems for positive systems under ℓ_1 and L_1 performance. Two classes of systems are considered. One is positive linear system and the other is positive fuzzy system. Some widely encountering factors such as time-delay and uncertainties are taken into consideration. The investigation of positive systems is more complicated than general systems because positive systems are defined on cones rather than linear spaces. Some well-established results for general linear systems cannot be readily applied to positive systems and new approaches should be derived for positive systems. The detailed problems being investigated are listed as follows:

- For positive systems with or without uncertainty, how to establish conditions under which the positive systems are stable and satisfy the ℓ_1 and L_1 performance?
- Given a positive system, how to design state-feedback and output-feedback controllers such that the closed-loop system is positive, stable and satisfy the performance? Furthermore, how to design positive state-bounding observers and positive filters to estimate the system state and output?

1.3 Literature Review

1.3.1 Positive Systems

Positive systems can be found in different application fields such as physical, engineering and social sciences [1, 2]. The variables of such systems take non-negative values at all times since they usually denote the concentrations or amounts of material in application fields. In other words, such systems involve quantities which are naturally non-negative.

Positive systems have special structures and possess many unique features. Therefore, many new problems appear and some previous approach used for general systems are no longer applicable to positive systems. Positive systems theory can be traced back to the study of nonnegative matrices, known as the famous Perron–Frobenius Theorem [3]. The pioneering work is due to Luenberger, who proposed a system-theoretic approach to positive systems in [4]. Since then, many results have been reported in the literature [5–11]. To name a few, necessary and sufficient conditions for positive realizability by means of convex analysis were derived in [12]. A positive state-space representation of a given transfer function was characterized by Farina and Benvenuti in [13]. Stability theory for nonnegative and compartmental dynamic systems with time delay was investigated in [14–19]. As for the results on 2-D positive systems, we refer readers to [8, 20, 21]. Necessary and sufficient conditions were provided to solve the stabilization problem of positive systems in [22, 23]. Some results on the model reduction problem for positive systems can be found in [24, 25].

Then, the following three topics for positive systems will be reviewed: stabilization, observer and filter synthesis.

Stabilization

As is well known, stability and stabilization problems are the basic issues for dynamic systems in control theory. It should be noted that for positive systems, not only the stability of the closed-loop system is required, but its positivity should be guaranteed. In recent years, a large amount of effort has been devoted to deriving stability criteria and stabilization results for positive systems. In [23], a formula for the real stability radius of uncertain positive continuous-time systems is established. In [26], a method is proposed for the state-feedback stabilization of positive linear systems based on Gersgorin's theorem. Moreover, it is pointed out that the stabilization problems for positive systems can be seen as the quadratic programming problems with constraints. In [27], the stabilization of equilibrium points of positive linear systems which are in the interior of the first orthant is considered. In [14], some necessary and sufficient stability criteria are proposed for positive systems with constant delays by means of linear Lyapunov–Krasovskii functionals. In [17], necessary and sufficient conditions are established for asymptotic stability of discrete-time positive systems with bounded time-varying delays. In [28], the stability problem is investigated for continuous-time positive systems with time-varying delays. It turns out that system stability is independent of the constant or time-varying delays. The stability analysis for switched positive systems is studied in [7, 29]. The internal stability of positive 2D systems is analyzed in [30]. The asymptotic stability for positive 2D systems with delays described by Roesser model, 2D Fornasini–Marchesini model and the general model is investigated in [31, 32].

As mentioned previously, in the synthesis of a control system, we should select appropriate control strategy and performance specifications according to practical situations. That is, despite of stability, suitable performance specification is also an important requirement in control system design. There are many performance specifications to measure the dynamic behavior of a control system, such as H_∞ performance, H_2 performance and so on. Among various performance specifications, the H_∞ performance characterizes the transfer ability of a control system to an energy bounded signal. It is a good measure for the robustness of a dynamic system and is often used to evaluate the attenuation ability to noise or model uncertainty. The H_∞ control problem is investigated for linear continuous-time and discrete-time positive systems in [33]. In [34], the problem of H_∞ control with $\mathcal{D}$-stability constraint is considered for a class of switched positive linear systems and the obtained conditions are formulated in terms of linear matrix inequalities. Besides the H_∞ norm, there are many other frequently used performance measures based on the ℓ_2 signal space [35]. However, it is noted that these performance measures induced by ℓ_2 signals are not very natural to describe some of the features of practical physical systems.

In some situations, for instance, if the variables denote the amount of material or the number of animal in a species, 1-norm can provide a more useful and natural description for positive systems because 1-norm gives the sum of the values of the components. On the other hand, different from many previous results based on quadratic Lyapunov functions, some results based on linear Lyapunov functions have emerged [1, 36, 37]. The motivation for using a linear Lyapunov function is that the state of a positive system is nonnegative and hence a linear Lyapunov function serves as a valid candidate. Fortunately, the ℓ_1 and L_1-induced indices fit well into the newly introduced linear Lyapunov function. Therefore, in recent years, many researchers are devoted to investigating controller synthesis based on ℓ_1 and L_1-induced norm. In [38], the L_1-optimal feedback controller is designed for positive systems with given weighing vectors. In [39], a novel characterization for stability and the ℓ_1-induced performance is proposed. Based on the characterization, a necessary and sufficient condition for the existence of desired controllers is derived with the use of linear Lyapunov function. In [40], the L_1-induced output-feedback controller is designed for continuous-time positive systems with interval uncertainties. In [41], the stability analysis and control of uncertain positive systems is considered with the L_1-induced norm and L_∞-induced norm using linear Lyapunov functions and dissipativity theory with linear supply rates.

Observer

Observer problems arise out of the need to estimate unmeasurable system variables. Due to its practical significance in the field of signal processing and control applications, the observer problem for general system has received great attention. The Luenberger-type observer design problem for general linear systems has been completely solved in [42]. An observer for nonlinear systems is constructed under rather general technical assumptions in [43]. The problem of H_∞ observer design is considered for a class of uncertain linear discrete-time systems with delayed state and parameter uncertainties in [44]. Due to the positivity of the system state, a valid estimate is required to be nonnegative for positive systems. However, it is noted that previous approach developed for general systems may not be applicable to the case of positive systems, since the estimated state cannot be guaranteed to be always nonnegative, that is, the state estimates may escape from the nonnegative orthant.

Recently, many results which can guarantee the positivity of the observer have been reported in [45–47]. In [48], positive observer is designed for compartmental systems through the structural decomposition of system matrices. Two approaches have been proposed to study the existence of positive observers for positive linear systems in [49]. One relies on a coordinates transformation and the other employs the theory of positive realization. In [50], linear time-invariant exponentially stable systems with additive disturbances are first transformed through a change of coordinates into cooperative systems and then time-varying exponentially stable interval observers are constructed. The method is applicable to general linear time-invariant systems, but may be difficult to incorporate performance requirements. Positive linear observers are designed for positive linear systems by coordinate transformation and positive realization in [49]. Unfortunately, these approaches lead to some constraints on the structure of positive observers and are not applicable to positive systems with parameter uncertainties. In [10], positive observers are designed for positive linear systems with interval uncertainties. In detail, necessary and sufficient conditions for the existence of positive observers with general structure are established, and the derived conditions are expressed in terms of LMIs.

It is noted that in the aforementioned works, the system state can only be estimated in an asymptotic way with conventional observers. However, the transient behavior of positive systems need to be known under certain circumstances [51, 52]. For example, in the wastewater treatment processes [51], the quantities of the pollutant in each compartment is required to be estimated during the whole process. Consequently, it is meaningful to design new observers which can give the information of the transient state of positive systems and many researchers are devoted to designing state-bounding positive observers for positive systems in recent years. For example, a pair of positive observers with state-bounding feature is proposed to estimate the state of positive systems at all times in [53].

In addition, due to the positivity of the state of positive systems, a linear Lyapunov function can be chosen as a valid candidate. This motivates the use of linear Lyapunov functions. In recent years, some researchers are devoted to investigating positive systems with linear Lyapunov functions [1, 36–41, 54]. By using a linear Lyapunov function, a novel method can be derived to address the observer design problem for positive systems. On the other hand, as mentioned previously, 1-norm can provide a more useful and natural description for positive systems. Therefore, ℓ_1 and L_1-induced performance can be used as appropriate characterizations for positive systems. The problem of L_1-induced performance based state-bounding observer design is studied for interval positive systems with their positivity preserved in the observer in [55]. In this paper, necessary and sufficient conditions are derived through linear Lyapunov functions and the results obtained are expressed in terms of linear programming problems.

Filtering

As a counterpart of the control problem, the filtering problem is to estimate the unmeasurable system variables from the measured output. In the area of control applications and signal processing, it has been a fundamental problem and has received much attention in recent years [56]. Among numerous filtering techniques, Kalman filtering is a classical one. A typical application of Kalman filtering is for navigation of aircraft. Due to the wide application of Kalman filtering, it have been investigated by many researchers [57–60]. Kalman filtering assumes that the exogenous input signals are stationary Gaussian noises and the properties of the external noises is required to be known. However, this requirement is difficult to satisfy in practice. Compared to Kalman filtering, H_∞ filtering assumes that the input signals are energy bounded and does not need the exact properties of the external noises. Furthermore, H_∞ filter is more robust against the uncertainties in systems. Therefore, the H_∞ filtering problem has received much attention. In [61], both full- and reduced-order H_∞ filters are designed for linear discrete-time uncertain systems via delay-dependent approach. In [62], the problem of H_∞ filters is addressed for continuous-time linear systems with stochastic uncertainties.

The filtering problem for positive systems has received relatively little attention and still remains challenging. Similar to the problem of observer design for positive systems, traditional approaches developed for the general systems cannot be directly applied to positive systems. The reason is that these methods cannot guarantee the positivity of the filter. In [9], the reduced-order H_∞ positive filtering problem is addressed for positive discrete-time systems and new approaches are presented to the filtering problem for positive systems with positivity preserved. As stated previously, ℓ_1 and L_1-induced performance can provide more natural descriptions and serve as appropriate characterizations for positive systems. Therefore, in recent years, some researchers have done some work on the filtering problem for positive systems under the ℓ_1 and L_1-induced performance. For example, in [63], the positive filtering problem is studied for positive continuous-time systems under the L_1-induced performance. Specifically speaking, a pair of positive filters with error-bounding feature is proposed to estimate the output of positive systems, that is, the information about the transient output can be given under such a framework.

1.3.2 Fuzzy Positive Systems

As is well known, physical systems and processes are nonlinear to a certain degree, which bring difficulties to the analysis and synthesis of dynamic systems. Therefore, many researchers have been devoted to finding an effective method to model and control the nonlinear systems in the past few decades. Among different methods, fuzzy modeling and control of complex nonlinear systems has received much attention. Fuzzy systems can be found in different applications such as control engineering, signal processing, decision making, robotics, and so on [64–67]. After the fuzzy set theory is introduced by Zadeh in [68–72], many contributions on fuzzy systems and control can be found in [64, 73–75]. Generally speaking, fuzzy system is characterized by a set of linguistic statements based on expert knowledge. The expert knowledge is usually in the form of rules. Usually, the If-Then rules are used to define fuzzy systems, which are of the following form: If *antecedent proposition,* then *consequent proposition*. In various rule-based fuzzy models, the T–S fuzzy model is one of the most commonly used types. T–S fuzzy model formulates the nonlinear systems into a framework consisting of local models connected by membership functions. That is, through the modeling approach, a nonlinear system can be described in terms of a set of local linear models. As a result, some research approaches applicable to linear systems can be used for nonlinear systems [76–78].

Many stability analysis results of T–S fuzzy systems have been derived based on the Lyapunov function method [78–80]. Some of them are obtained with the common quadratic Lyapunov functions [78, 79, 81]. However, it is shown that the results derived by common quadratic Lyapunov functions are conservative [82, 83]. To relax the conservatism, other types of Lyapunov functions have been used by many researchers. Less conservative stability results based on piecewise quadratic Lyapunov functions are developed for T–S fuzzy systems [84, 85]. Besides using piecewise quadratic Lyapunov functions, stability results based on fuzzy Lyapunov functions are presented in [76, 86, 87]. It should be mentioned that common quadratic Lyapunov functions can be seen as a special case of the more general piecewise quadratic and fuzzy Lyapunov functions. Meanwhile, based on the stability results, stabilization and other synthesis problems for T–S fuzzy systems are investigated in [88–92].

Despite the success, there has been little research on nonlinear positive systems. Therefore, some attention has been devoted to studying the problems of nonlinear positive systems in recent years. Inspired by [93], multiple Lyapunov functions are used to obtain sufficient conditions of asymptotic stability for positive T–S fuzzy discrete-time linear systems in [94]. However, it is noted that in [94], the stability and stabilization of nonlinear positive systems is investigated based on quadratic Lyapunov functions. As stated previously, positive systems possess some special characteristics, and thus it is natural to expect that new methods may be used to obtain some desirable results for fuzzy positive systems. From the existing results on positive systems, it is known that linear Lyapunov function captures the nature of positivity and by using the linear Lyapunov function, necessary and sufficient stability conditions can be obtained for positive linear systems. As a result, some novel results for positive fuzzy systems have appeared, which are derived based on linear Lyapunov functions. For example, based on a novel linear copositive Lyapunov function, the stability and constrained control are addressed for fuzzy positive systems with delays in [95]. In [96], with linear Lyapunov functions, three approaches are proposed to investigate the stability of discrete-time fuzzy positive systems with constant time delays. In [97], the problem of ℓ_1-induced filtering is investigated for positive T–S fuzzy systems with the positivity preserved in the filters.

1.4 Mathematical Preliminaries

1.4.1 Definitions and Lemmas on Positive Linear Systems

In this subsection, the basic definitions on positive linear systems are presented. Then, the positivity and stability characterization for positive systems are provided. Moreover, some other important lemmas about positive systems are given.

Definition 1.1 (*Discrete-time Positive Linear Systems*)
Consider a discrete-time linear system:

$$\begin{cases} x\left(k+1\right) = Ax\left(k\right) + B_w w\left(k\right), \\ y\left(k\right) = Cx\left(k\right) + D_w w\left(k\right), \\ x\left(0\right) = x_0, \end{cases} \tag{1.1}$$

where $x(k) \in \mathbb{R}^n$, $w(k) \in \mathbb{R}^m$ and $y(k) \in \mathbb{R}^q$ are the system state, input and output, respectively. A, B_w, C and $D_w$are real constant matrices with appropriate dimensions. System (1.1) is said to be a discrete-time positive linear system if for all $x_0 \succeq 0$ and all inputs $w(k) \succeq 0$, we have $x(k) \succeq 0$ and $y(k) \succeq 0$ for $k \geq 0$.

Definition 1.2 (*Continuous-time Positive Linear Systems*)

Consider a continuous-time linear system:

$$\begin{cases} \dot{x}(t) = Ax(t) + B_w w(t), \\ y(t) = Cx(t) + D_w w(t), \\ x(0) = x_0, \end{cases} \tag{1.2}$$

where $x(t) \in \mathbb{R}^n$, $w(t) \in \mathbb{R}^m$ and $y(t) \in \mathbb{R}^q$ are the system state, input and output, respectively. A, B_w, C and D_w are real constant matrices with appropriate dimensions. System (1.2) is said to be a continuous-time positive linear system if for all $x_0 \succeq 0$ and all inputs $w(t) \succeq 0$, we have $x(t) \succeq 0$ and $y(t) \succeq 0$ for $t > 0$.

Definition 1.3 For a matrix $A \in \mathbb{R}^{m\times n}$, $A \geq\geq 0$ (respectively, $A \gg 0$) means that all its elements are nonnegative, that is, $\forall (i, j)$, $a_{ij} \geq 0$ (respectively, $a_{ij} > 0$), where a_{ij} denotes the element located at the ith row and the jth column.

Definition 1.4 For a matrix $A \in \mathbb{R}^{n\times n}$, $A \in \mathbb{R}^{n\times n}$ is called Metzler, if all its off-diagonal elements are nonnegative, that is, $\forall (i, j)$, $i \neq j$, $a_{ij} \geq 0$, where a_{ij} denotes the element located at the ith row and the jth column.

Lemma 1.1 (Positivity Characterization [22])

1. *System (1.1) is a discrete-time positive linear system if and only if*

$$A \geq\geq 0,\ B_w \geq\geq 0,\ C \geq\geq 0,\ D_w \geq\geq 0.$$

2. *System (1.2) is a continuous-time positive linear system if and only if*

$$A \text{ is Metzler},\ B_w \geq\geq 0,\ C \geq\geq 0,\ D_w \geq\geq 0.$$

Lemma 1.2 (Stability Characterization [37])

1. *The discrete-time positive linear system in (1.1) is asymptotically stable if and only if there exists a vector $p \geq\geq 0$ (or $p \gg 0$) satisfying*

$$p^T(A - I) \ll 0.$$

2. *The continuous-time positive linear system in (1.2) is asymptotically stable if and only if there exists a vector $p \geq\geq 0$ (or $p \gg 0$) satisfying*

$$p^T A \ll 0.$$

In the following, we introduce some other important definitions and lemmas which will be used in the sequel.

Definition 1.5 For a vector $x = [x_1, \ldots, x_n]^T \in \mathbb{R}^n$, it is called sparse if its ℓ_0-norm is small compared to the dimension of the vector, where ℓ_0-norm of x is defined as

$$\|x\|_0 = \sum_{i=1}^{n} |\text{sign}(x_i)|.$$

Lemma 1.3 ([98]) *For two nonnegative matrices $A_1, A_2 \in \mathbb{R}^{n\times n}$, if $A_1 \geq\geq A_2$, then $\rho(A_1) \geq \rho(A_2)$.*

Lemma 1.4 ([98]) *For Metzler matrix A, $-A^{-1} \geq\geq 0$ if and only if $\alpha(A) < 0$.*

Lemma 1.5 ([40]) *For two Metzler matrices $A_1, A_2 \in \mathbb{R}^{n\times n}$, if $A_1 \geq\geq A_2$, then $\alpha(A_1) \geq \alpha(A_2)$. Moreover, if $\alpha(A_1) < 0$, then $-A_1^{-1} \geq\geq -A_2^{-1}$.*

Lemma 1.6 ([40]) *For $0 \leq\leq M^* \leq\leq M$ and $0 \leq\leq N^* \leq\leq N$ of compatible dimensions, we have*

$$0 \leq\leq M^*N^* \leq\leq MN.$$

Lemma 1.7 ([99]) *The convex envelope of the function $f = \|x\|_0 = \sum_{i=1}^{n} |\text{sign}(x_i)|$ on $X = \{x \in \mathbb{R}^n | \|x\|_\infty \leq 1\}$ is $f_{env}(x) = \|x\|_1 = \sum_{i=1}^{n} |x_i(t)|$.*

1.4.2 Definitions on Fuzzy Positive Systems

In this subsection, we briefly introduce the definitions of fuzzy positive systems. Moreover, the positivity characterization of fuzzy positive systems is given.

Definition 1.6 (*Fuzzy Positive Systems*)

Consider the following fuzzy system described by the ith rule as follows:

Model Rule i: IF $\theta_1(k)$ is M_{i1} and $\theta_2(k)$ is M_{i2} and ... and $\theta_g(k)$ is M_{ig}, THEN

$$\begin{cases} x(k+1) = A_i x(k) + B_{wi} w(k), \\ y(k) = C_i x(k) + D_{wi} w(k), \end{cases} \tag{1.3}$$

where $x(k) \in \mathbb{R}^n$, $w(k) \in \mathbb{R}^m$ and $y(k) \in \mathbb{R}^q$ denote the system state, disturbance input and output, respectively. $i = 1, 2, \ldots, r$, is the number of rules and $\theta_1(k), \theta_2(k), \ldots, \theta_g(k)$ are the premise variables. M_{ie} $(i = 1, 2, \ldots, r; e = 1, 2, \ldots, g)$ represents the fuzzy sets. Then, we have the final fuzzy system:

$$\begin{cases} x(k+1) = \sum_{i=1}^{r} h_i(\theta(k))(A_i x(k) + B_{wi} w(k)), \\ y(k) = \sum_{i=1}^{r} h_i(\theta(k))(C_i x(k) + D_{wi} w(k)), \end{cases} \tag{1.4}$$

where

$$h_i(\theta(k)) = \mu_i(\theta(k)) / \sum_{i=1}^{r} \mu_i(\theta(k)), \quad \mu_i(\theta(k)) = \prod_{e=1}^{g} M_{ie}(\theta_e(k)),$$

and $M_{ie}(\theta_e(k)) \in [0, 1]$ represents the grade of membership of $\theta_e(k)$ in M_{ie}. For all k we have

$$\sum_{i=1}^{r} h_i(\theta(k)) = 1, \ h_i(\theta(k)) \geq 0, \qquad i = 1, 2, \ldots, r.$$

System (1.4) is a discrete-time positive fuzzy system if for all $x(0) \geq\geq 0$ and input $w(k) \geq\geq 0$, we have $x(k) \geq\geq 0$ and $y(k) \geq\geq 0$ for $k \in \mathbb{N}$.

Next, the important positivity characterization is introduced.

Lemma 1.8 (Positivity Characterization [94])

The discrete-time system (1.4) is positive if and only if

$$A_i \geq\geq 0, \ B_{wi} \geq\geq 0, \ C_i \geq\geq 0, \ D_{wi} \geq\geq 0, \ i = 1, 2, \ldots, r.$$

1.4.3 ℓ_1 and L_1 Performance

In this subsection, the ℓ_1 and L_1 performance of positive systems is introduced.

ℓ_1-Induced Performance

For a stable discrete-time positive system (1.1), we say that it has ℓ_1-induced performance at the level γ if, under zero initial conditions, for all nonzero $w \in \ell_1$,

$$\|y\|_{\ell_1} < \gamma \|w\|_{\ell_1}, \tag{1.5}$$

or equivalently,

$$\sum_{k=0}^{\infty} \|y(k)\|_1 < \gamma \sum_{k=0}^{\infty} \|w(k)\|_1, \tag{1.6}$$

where $\gamma > 0$ is a given performance level.

L_1-Induced Performance

For a stable continuous-time positive system (1.2), we say that it has L_1-induced performance at the level γ if, under zero initial conditions, for all nonzero $w \in L_1$,

$$\|y\|_{L_1} < \gamma \|w\|_{L_1}, \tag{1.7}$$

or equivalently,

$$\int_0^{\infty} \|y(t)\|_1 dt < \gamma \int_0^{\infty} \|w(t)\|_1 dt, \tag{1.8}$$

where $\gamma > 0$ is a given scalar.

1.5 Thesis Outline

- Chapter 2 deals with the problem of ℓ_1-induced state-feedback controller design for discrete-time positive systems. First, an analytical method to compute the exact value of ℓ_1-induced norm is presented. Then, a novel characterization for stability and ℓ_1-induced performance is proposed. Based on the characterization, a necessary and sufficient condition for the existence of desired controllers is derived, and an iterative convex optimization approach is developed to solve the condition. In addition, the synthesis of the state-feedback controller for SIMO positive systems is investigated. For this special case, an analytic solution is established to show how the optimal ℓ_1-induced controller can be designed, and some links to the spectral radius of the closed-loop systems are provided. The theoretical results are illustrated through a numerical example.
- Chapter 3 is concerned with the design of L_1-induced output-feedback controller for continuous-time positive systems with interval uncertainties. An analytical method is first presented to compute the exact value of L_1-induced norm. Then, a necessary and sufficient condition for the stability and L_1-induced performance of positive linear systems is proposed in terms of linear inequalities. Based on this, conditions for the existence of robust static output-feedback controllers are established and an iterative convex optimization approach is developed to solve the conditions. When $C = I$, the static output-feedback controller synthesis problem is reduced to a state-feedback controller synthesis problem. Analytical design approach is put forward to investigate the controller synthesis for interval continuous positive systems with $C = I$. In addition, this chapter also addresses the design of L_1-induced sparse state-feedback controller for continuous-time positive systems with interval uncertainties. That is, the total number of all the nonzero elements of the controller gain is to be minimized, while satisfying a guaranteed level of L_1-induced performance. Moreover, a dynamic output-feedback controller is designed for interval uncertain positive systems. Several illustrative examples are provided to show the effectiveness and applicability of the theoretical results.
- Chapter 4 treats the design problem of positive state-bounding observers for interval positive systems under the L_1-induced performance. To estimate the state of positive systems at all times, a pair of state-bounding positive observers is designed. A novel characterization is first proposed under which the augmented system is stable and satisfies the L_1-induced performance. Necessary and sufficient conditions are then presented to design observers. The results obtained in this chapter are expressed in terms of linear programming problems, and can be easily solved by standard software. The effectiveness of the derived design procedures is demonstrated through a numerical example.
- Chapter 5 addresses the positive filtering problem for positive continuous-time systems under the L_1-induced performance. A pair of positive filters with error-bounding feature is proposed to estimate the output of positive systems. A novel characterization is first obtained to ensure the filtering error system asymptotically stable with a prescribed L_1-induced performance. Then, necessary and sufficient

conditions for the existence of required filters are presented, and the obtained results are expressed in terms of linear programming problems, which can be easily checked by standard software. A numerical example are used to illustrate the developed theories.

- Chapter 6 investigates the ℓ_1-induced controller synthesis and positive filtering problems for positive Takagi–Sugeno (T–S) fuzzy systems. Novel performance characterization of the positive fuzzy systems is established. Based on the characterization, sufficient conditions are established for the existence of state-feedback controller. An iterative convex optimization algorithm is developed to solve the design conditions. Moreover, to estimate the output of positive T–S fuzzy systems, error-bounding positive filters are constructed. A new performance characterization is first established to guarantee the asymptotic stability of the filtering error system with the ℓ_1-induced performance. Then, sufficient conditions expressed by linear programming problems are derived to design the required filters. Finally, Two examples are presented to demonstrate the effectiveness of the proposed approach.

1.6 Contributions

The major contributions of the thesis are summarized as follows:

- For discrete-time positive systems, a method to compute the exact value of ℓ_1-induced norm is proposed. An ℓ_1-induced performance index is explicitly presented and a characterization is developed under which the positive linear system is asymptotically stable and satisfies the ℓ_1-induced performance. Similarly, for continuous-time positive systems, the L_1-induced performance is introduced for continuous-time positive systems and the exact value of L_1-induced norm is computed. A necessary and sufficient condition for stability and L_1-induced performance of positive linear systems is proposed in terms of linear inequalities.
- For discrete-time positive systems, necessary and sufficient conditions for the existence of the desired ℓ_1-induced controller are derived. For continuous-time positive systems with interval uncertainties, conditions for the existence of robust L_1-induced output-feedback controllers are established. Moreover, the L_1-induced sparse controller is designed for continuous-time positive systems with interval uncertainties. In addition, iterative convex optimization approaches are presented to solve the conditions.
- The synthesis of the ℓ_1-induced and L_1-induced state-feedback controllers for single-input multiple-output (SIMO) positive systems is investigated. For the special case, analytic solutions are established to show how the optimal ℓ_1-induced and L_1-induced controllers can be designed. The results can be further extended to the multiple-input-multiple-output (MIMO) case by employing structured controllers.

- Different from conventional observers which only provide an estimate of the system state in an asymptotic way, a pair of positive observers with state-bounding feature is proposed to estimate the state of positive systems at all times. Similarly, necessary and sufficient conditions are presented to design a pair of error-bounding filters under the L_1-induced performance. The results obtained are expressed in terms of linear programming problems, and can be easily solved by standard software. Moreover, we extend the filtering design results to positive Takagi–Sugeno (T–S) fuzzy systems. Sufficient conditions expressed by linear programming problems are derived to design the required filters.

References

1. Haddad WM, Chellaboina VS (2005) Stability and dissipativity theory for nonnegative dynamical systems: a unified analysis framework for biological and physiological systems. Nonlinear Anal: Real World Appl 6(1):35–65
2. Muratori S, Rinaldi S (1990) Equilibria, stability, and reachability of Leslie systems with nonnegative inputs. IEEE Trans Autom Control 35(9):1065–1068
3. Berman A, Plemmons RJ (1994) Nonnegative matrices in the mathematical sciences. SIAM, Philadephia
4. Luenberger D (1979) Introduction to dynamic systems: theory, models, and applications. Wiley, New York
5. Rami MA, Schönlein M, Jordan J (2013) Estimation of linear positive systems with unknown time-varying delays. Eur J Control 19(3):179–187
6. Feng J, Lam J, Li P et al (2011) Decay rate constrained stabilization of positive systems using static output feedback. Int J Robust Nonlinear Control 21(1):44–54
7. Fornasini E, Valcher ME (2010) Linear copositive Lyapunov functions for continuous-time positive switched systems. IEEE Trans Autom Control 55:1933–1937
8. Kaczorek T (2002) Positive 1D and 2D Systems. Springer, Berlin
9. Li P, Lam J, Shu Z et al (2010) H_∞ positive filtering for positive linear discrete-time systems: an augmentation approach. IEEE Trans Autom Control 55(10):2337–2342
10. Shu Z, Lam J, Gao H et al (2008) Positive observers and dynamic output-feedback controllers for interval positive linear systems. IEEE Trans Circuits Syst (I) 55(10):3209–3222
11. Valcher ME (2009) Reachability properties of continuous-time positive systems. IEEE Trans Autom Control 54:1586–1590
12. Farina L (1996) On the existence of a positive realization. Syst Control Lett 28:219–226
13. Benvenuti L, Farina L (2004) A tutorial on the positive realization problem. IEEE Trans Autom Control 49(5):651–664
14. Haddad WM, Chellaboina VS (2004) Stability theory for nonnegative and compartmental dynamical systems with time delay. Syst Control Lett 51(5):355–361
15. Haddad WM, Chellaboina VS, Rajpurohit T (2004) Dissipativity theory for nonnegative and compartmental dynamical systems with time delay. IEEE Trans Autom Control 49(5):747–751
16. Liu X, Lam J (2013) Relationships between asymptotic stability and exponential stability of positive delay systems. Int J Gen Syst 42(2):224–238
17. Liu X, Yu W, Wang L et al (2009) Stability analysis of positive systems with bounded time-varying delays. IEEE Trans Circuits Syst (II) 56(7):600–604
18. Shen J, Lam J (2014) L_∞-gain analysis for positive systems with distributed delays. Automatica 50(1):175–179
19. Wu L, Lam J, Shu Z et al (2009) On stability and stabilizability of positive delay systems. Asian J Control 11(2):226–234

20. Kurek JE (2002) Stability of positive 2-D system described by the Roesser model. IEEE Trans Circuits Syst (I) 49:531–533
21. Valcher ME, Fornasini E (1995) State models and asymptotic behaviour of 2D positive systems. IMA J Math Control Inf 12(1):17–36
22. Farina L, Rinaldi S (2011) Positive linear systems: theory and applications. Wiley, New York
23. Son NK, Hinrichsen D (1996) Robust stability of positive continuous-time systems. Numer Funct Anal Opt 17:649–659
24. Feng J, Lam J, Shu Z et al (2010) Internal positivity preserved model reduction. Int J Control 83(3):575–584
25. Li P, Lam J, Wang Z et al (2011) Positivity-preserving H_∞ model reduction for positive systems. Automatica 47(7):1504–1511
26. Kaczorek T (1999) Stabilization of positive linear systems by state feedback. Pomiary, Automatyka, Kontrola 3:2–5
27. De Leenheer P, Aeyels D (2001) Stabilization of positive linear systems. Syst Control Lett 44(4):259–271
28. Liu X, Yu W, Wang L et al (2010) Stability analysis for continuous-time positive systems with time-varying delays. IEEE Trans Autom Control 55(4):1024–1028
29. Gurvits L, Shorten R, Mason O (2007) On the stability of switched positive linear systems. IEEE Trans Autom Control 52(6):1099–1103
30. Valcher ME (1997) On the internal stability and asymptotic behavior of 2-D positive systems. IEEE Trans Circuits Syst (I) 44(7):602–613
31. Kaczorek T (2009) Asymptotic stability of positive 2D linear systems with delays. Bull Polish Acad Sci: Tech Sci 57(2):133–138
32. Kaczorek T (2009) LMI approach to stability of 2D positive systems. Multidimens Syst Signal Process 20(1):39–54
33. He Z, Wu L, Su X (2009) H_∞ control of linear positive systems: continuous-and discrete-time cases. Int J Innov Comput Inf Control 5(6):1747–1756
34. Tong Y, Zhang L, Basin M et al (2014) Weighted H_∞ control with D-stability constraint for switched positive linear systems. Int J Robust Nonlinear Control 24(4):758–774
35. Lu WW, Balas GJ (1988) A comparison between Hankel norms and induced system norms. IEEE Trans Autom Control 43(11):1658–1662
36. Rami MA, Tadeo F (2007) Controller synthesis for positive linear systems with bounded controls. IEEE Trans Circuits Syst (II) 54(2):151–155
37. Haddad WM, Chellaboina VS, Hui Q (2010) Nonnegative and compartmental dynamical systems. Princeton University Press, Princeton
38. Ebihara Y, Peaucelle D, Arzelier D (2012) Optimal L_1-controller synthesis for positive systems and its robustness properties. In: Proceedings of the 2012 American control conference
39. Chen X, Lam J, Li P et al (2013) ℓ_1-induced norm and controller synthesis of positive systems-induced norm and controller synthesis of positive systems. Automatica 49:1377–1385
40. Chen X, Lam J, Li P et al (2014) Output-feedback control for continuous-time interval positive systems under L_1 performance. Asian J Control 16(6):1592–1601
41. Briat C (2013) Robust stability and stabilization of uncertain linear positive systems via integral linear constraints: L_1-gain and L_∞-gain characterization. Int J Robust Nonlinear Control 23(17):1932–1954
42. David G (1971) An introduction to observers. IEEE Trans Autom Control 16(6):596–602
43. Gauthier JP, Hammouri H, Othman S (1992) A simple observer for nonlinear systems applications to bioreactors. IEEE Trans Autom Control 37(6):875–880
44. Wang Z, Huang B, Unbehauen H et al (1999) Robust H_∞ observer design of linear state delayed systems with parametric uncertainty: the discrete-time case. Automatica 35(6):1161–1167
45. Rami MA, Cheng CH, De Prada C (2008) Tight robust interval observers: an LP approach. In: Proceedings of the 47th IEEE conference on decision and control
46. Rami MA, Helmke U, Tadeo F (2007) Positive observation problem for linear time-delay positive systems. In: Mediterranean conference on control and automation

47. Rami MA, Helmke U, Tadeo F (2007) Positive observation problem for linear time-lag positive systems. In: Proceedings of the third IFAC symposium on system structure and control
48. Van Den Hof JM (1998) Positive linear observers for linear compartmental systems. SIAM J Control Optim 36(2):590–608
49. Back J, Astolfi A (2008) Design of positive linear observers for positive linear systems via coordinate transformations and positive realizations. SIAM J Control Optim 47(1):345–373
50. Mazenc F, Bernard O (2011) Interval observers for linear time-invariant systems with disturbances. Automatica 47(1):140–147
51. Lemesle V, Gouzé JL (2005) Hybrid bounded error observers for uncertain bioreactor models. Bioprocess Biosyst Eng 27(5):311–318
52. Moisan M, Bernard O, Gouzé JL et al (2009) Near optimal interval observers bundle for uncertain bioreactors. Automatica 45(1):291–295
53. Li P, Lam J (2012) Positive state-bounding observer for positive interval continuous-time systems with time delay. Int J Robust Nonlinear Control 22:1244–1257
54. Chen X, Lam J, Li P et al (2013) L_1-induced performance analysis and sparse controller synthesis for interval positive systems. In: Proceedings of the international conference of applied and engineering mathematics
55. Chen X, Lam J (2014) Positive state-bounding observer for interval positive systems under L_1 performance. In: Proceedings of the 33rd chinese control conference
56. Anderson BDO, Moore JB (2012) Optimal filtering. Courier Dover Publications, Mineola
57. Chui CK, Chen G (1999) Kalman filtering. Springer, Berlin
58. Grewal MS, Andrews AP (2011) Kalman filtering: theory and practice using MATLAB. Wiley, Canada
59. Haykin SS (2001) Kalman filtering and neural networks. Wiley Online Library, New York
60. Sinopoli B, Schenato L, Franceschetti M et al (2004) Kalman filtering with intermittent observations. IEEE Trans Autom Control 49(9):1453–1464
61. Gao H, Wang C (2004) A delay-dependent approach to robust H_∞ filtering for uncertain discrete-time state-delayed systems. IEEE Trans Signal Process 52(6):1631–1640
62. Gershon E, Limebeer DJN, Shaked U et al (2001) Robust H_∞ filtering of stationary continuous-time linear systems with stochastic uncertainties. IEEE Trans Autom Control 46(11):1788–1793
63. Chen X, Lam J, Li P (2014) Positive filtering for continuous-time positive systems under L_1 performance. Int J Control 87(9):1906–1913
64. Babuska R (1998) Fuzzy modeling for control. Kluwer Academic Publishers, Boston
65. Bellman RE, Zadeh LA (1970) Decision-making in a fuzzy environment. Manag Sci 17(4):B141
66. Dutta S (1993) Fuzzy logic applications: technological and strategic issues. IEEE Trans Eng Manag 40(3):237–254
67. Keller J, Krisnapuram R, Pal NR et al (2005) Fuzzy models and algorithms for pattern recognition and image processing. Springer, Berlin
68. Zadeh LA (1965) Fuzzy sets. Inf Control 8(3):338–353
69. Zadeh LA (1968) Fuzzy algorithms. Inf Control 12(2):94–102
70. Zadeh LA (1971) Similarity relations and fuzzy orderings. Inf Sci 3(2):177–200
71. Zadeh LA (1973) Outline of a new approach to the analysis of complex systems and decision processes. IEEE Trans Syst Man Cybern 1:28–44
72. Zadeh LA (1975) The concept of a linguistic variable and its application to approximate reasoning-I. Inf Sci 8(3):199–249
73. Lee CC (1990) Fuzzy logic in control systems: fuzzy logic controller. II. IEEE Trans Syst Man Cybern 20(2):419–435
74. Su X, Shi P, Wu L et al (2013) A novel control design on discrete-time Takagi-Sugeno fuzzy systems with time-varying delays. IEEE Trans Fuzzy Syst 21(4):655–671
75. Wang L (1999) A course in fuzzy systems. Prentice-Hall press, USA
76. Guerra TM, Vermeiren L (2004) LMI-based relaxed nonquadratic stabilization conditions for nonlinear systems in the Takagi-Sugeno's form. Automatica 40(5):823–829

77. Tanaka K, Ikeda T, Wang HO et al (1998) Fuzzy regulators and fuzzy observers: relaxed stability conditions and LMI-based designs. IEEE Trans Fuzzy Syst 6(2):250–265
78. Tuan HD, Apkarian P, Narikiyo T et al (2001) Parameterized linear matrix inequality techniques in fuzzy control system design. IEEE Trans Fuzzy Syst 9(2):324–332
79. Teixeira M, Assunção E, Avellar RG et al (2003) On relaxed LMI-based designs for fuzzy regulators and fuzzy observers. IEEE Trans Fuzzy Syst 11(5):613–623
80. Wang Z, Ho DWC, Liu X et al (2004) A note on the robust stability of uncertain stochastic fuzzy systems with time-delays. IEEE Trans Syst Man Cybern - Part A 34(4):570–576
81. Tanaka K, Sugeno M (1992) Stability analysis and design of fuzzy control systems. Fuzzy Sets Syst 45(2):135–156
82. Feng G (2004) Stability analysis of discrete-time fuzzy dynamic systems based on piecewise Lyapunov functions. IEEE Trans Fuzzy Syst 12(1):22–28
83. Johansson M, Rantzer A, Arzen K (1999) Piecewise quadratic stability of fuzzy systems. IEEE Trans Fuzzy Syst 7(6):713–722
84. Cao SG, Rees NW, Feng G (1996) Stability analysis and design for a class of continuous-time fuzzy control systems. Int J Control 64(6):1069–1087
85. Cao SG, Rees NW, Feng G (1997) Analysis and design for a class of complex control systems part II: fuzzy controller design. Automatica 33(6):1029–1039
86. Choi DJ, Park PG (2003) H_∞ state-feedback controller design for discrete-time fuzzy systems using fuzzy weighting-dependent Lyapunov functions. IEEE Trans Fuzzy Syst 11(2):271–278
87. Tanaka K, Hori T, Wang HO et al (2003) A multiple Lyapunov function approach to stabilization of fuzzy control systems. IEEE Trans Fuzzy Syst 11(4):582–589
88. Chen X, Lam J, Gao H et al (2013) Stability analysis and control design for 2-D fuzzy systems via basis-dependent Lyapunov functions. Multidimens Syst Signal Process 24(3):395–415
89. Jadbabaie A (1999) A reduction in conservatism in stability and L_2 gain analysis of Takagi-Sugeno fuzzy systems via linear matrix inequalities. In: Proceedings of 14th IFAC Triennial world congress. Beijing, China, pp 285–289
90. Tanaka K, Wang HO (2001) Fuzzy control systems design and analysis: a linear matrix inequality approach. Wiley, New York
91. Tanaka K, Hori T, Wang HO (2001) A fuzzy Lyapunov approach to fuzzy control system design. In: Proceedings of the American control conference. Arlington, VA, pp 4790–4795
92. Zhou S, Li T (2005) Robust stabilization for delayed discrete-time fuzzy systems via basis-dependent Lyapunov-Krasovskii function. Fuzzy Sets Syst 151(1):139–153
93. Benzaouia A, Mehdi D, El Hajjaji A et al (2007) Piecewise quadratic Lyapunov function for nonlinear systems with fuzzy static output feedback control. In: Proceedings of the European control conference, Kos, Greece
94. Benzaouia A, Hmamed A, Hajjaji EL, Hajjaji EL A (2010) Stabilization of controlled positive discrete-time T-S fuzzy systems by state feedback control. Int J Adap Control Signal Process 24:1091–1106
95. Mao Y, Zhang H, Dang C et al (2012) Stability analysis and constrained control of a class of fuzzy positive systems with delays using linear copositive Lyapunov functional. Circuits Syst Signal Process 31(5):1863–1875
96. Wu Y, Luo H, Zhang H et al (2013) Stability analysis of discrete-time fuzzy positive systems with time delays. J Intell Fuzzy Syst 25(4):893–905
97. Chen X, Lam J, Lam HK (2015) Positive filtering for positive Takagi-Sugeno fuzzy systems under ℓ_1 performance. Inf Sci 299:32–41
98. Bapat RB, Raghavan TES (1997) Nonnegative matrices and applications. Cambridge University Press, Cambridge
99. Fazel M (2002) Matrix rank minimization with applications. PhD thesis, Department of Electrical Engineering, Stanford University

Chapter 2
ℓ_1-Induced Controller Design for Positive Systems

In this chapter, the problem of ℓ_1-induced state-feedback controller design is investigated for discrete-time positive systems based on linear Lyapunov functions. For a given positive linear system, the objective is to design a state-feedback controller such that the closed-loop system is positive, asymptotically stable and satisfies a prescribed ℓ_1-induced performance. In detail, the exact value of ℓ_1-induced norm is firstly computed with the use of an analytical method. A novel stability and ℓ_1-induced performance characterization is then proposed. Based on such a characterization, necessary and sufficient conditions are derived for the existence of desired controllers. To solve the conditions, a corresponding iterative convex optimization algorithm is developed. Secondly, the controller synthesis problem for single-input multiple-output (SIMO) positive systems is investigated. Specifically speaking, an analytical method is established to obtain the optimal ℓ_1-induced controller for the special case. Moreover, some links to the spectral radius of the closed-loop systems are provided. Finally, the effectiveness of the obtained theoretical results in this chapter are illustrated through a numerical example.

The organization of this chapter is as follows. Section 2.1 formulates the ℓ_1-induced state-feedback controller synthesis problem for positive systems. In Sect. 2.2, a method to compute the exact value of ℓ_1-induced norm is first proposed. An equivalent characterization is then developed under which the positive linear system is asymptotically stable and satisfies the ℓ_1-induced performance. In Sect. 2.3, the controller design method for positive systems is put forward and an iterative algorithm is developed to solve the derived conditions. A novel condition for the existence of the state-feedback controller for SIMO positive systems is obtained in Sect. 2.4. Section 2.5 presents an illustrative example. A summary is given in Sect. 2.6.

X. Chen, *Analysis and Synthesis of Positive Systems Under ℓ_1 and L_1 Performance*, Springer Theses, DOI 10.1007/978-981-10-2227-2_2

2.1 Problem Formulation and Preliminaries

Consider a discrete-time positive linear system:

$$\begin{cases} x(k+1) = Ax(k) + Bu(k) + B_w w(k), \\ y(k) = Cx(k) + Du(k) + D_w w(k), \end{cases} \tag{2.1}$$

where $x(k) \in \mathbb{R}^n$, $u(k) \in \mathbb{R}^l$, $w(k) \in \mathbb{R}^m$ and $y(k) \in \mathbb{R}^r$ are the system state, control input, disturbance input and output, respectively; $A \in \mathbb{R}^{n\times n}$, $B \in \mathbb{R}^{n\times l}$, $B_w \in \mathbb{R}^{n\times m}$, $C \in \mathbb{R}^{r\times n}$, $D \in \mathbb{R}^{r\times l}$ and $D_w \in \mathbb{R}^{r\times m}$ are nonnegative.

In this chapter, the state-feedback controller will be designed in the form of

$$u(k) = Kx(k). \tag{2.2}$$

Then, the following closed-loop system is obtained:

$$\begin{cases} x(k+1) = (A+BK)x(k) + B_w w(k), \\ y(k) = (C+DK)x(k) + D_w w(k). \end{cases} \tag{2.3}$$

For the system in (2.3), we say that it has ℓ_1-induced performance at the level γ if, under zero initial conditions, for all nonzero $w \in \ell_1$,

$$\|y\|_{\ell_1} < \gamma \|w\|_{\ell_1}, \tag{2.4}$$

or equivalently,

$$\sum_{k=0}^{\infty} \|y(k)\|_1 < \gamma \sum_{k=0}^{\infty} \|w(k)\|_1, \tag{2.5}$$

where $\gamma > 0$ is a given performance level.

In this investigation, apart from the ℓ_1-induced performance in (2.4), positivity is required to be preserved. Then, the problem to be addressed in this chapter is described as follows.

Problem PPL1CD (Positivity-Preserving ℓ_1-Induced Controller Design): Given the positive system (2.1), the control objective is to find a controller $u(k) = Kx(k)$ such that the closed-loop system (2.3) is positive, asymptotically stable, and satisfies the ℓ_1-induced performance in (2.4) under zero initial conditions.

2.2 Performance Analysis

Consider a discrete-time positive linear system:

$$\begin{cases} x(k+1) = Ax(k) + B_w w(k), \quad x(0) = x_0, \\ y(k) = Cx(k) + D_w w(k), \end{cases} \tag{2.6}$$

where $x(k) \in \mathbb{R}^n$, $w(k) \in \mathbb{R}^m$ and $y(k) \in \mathbb{R}^r$ are the system state, input, and output, respectively; A, B_w, C and D_w are nonnegative matrices. In this section, we first propose a method to compute the exact value of ℓ_1-induced norm for system (2.6). Then a characterization is proposed under which the positive linear system is asymptotically stable and satisfies the performance in (2.4).

2.2.1 Exact Computation of ℓ_1-Induced Norm

In the following theorem, we establish an analytical characterization through which the value of ℓ_1-induced norm of system (2.6) is computed directly.

Theorem 2.1 *For an asymptotically stable positive linear system given in (2.6), denote the system operator $G : \ell_1 \to \ell_1$ for input w to output y. Then, the exact value of the ℓ_1-induced norm is given by*

$$\|G\|_{(\ell_1,\ell_1)} = \|C(I - A)^{-1}B_w + D_w\|_1. \tag{2.7}$$

Proof From system (2.6), we know that

$$\begin{aligned}
y(0) &= Cx(0) + D_w w(0),\\
y(1) &= CAx(0) + CB_w w(0) + D_w w(1),\\
y(2) &= CA^2x(0) + CB_w w(1) + CAB_w w(0) + D_w w(2),\\
&\vdots\\
y(k) &= CA^k x(0) + C\sum_{m=1}^{k} A^{k-m}B_w w(m-1) + D_w w(k).
\end{aligned}$$

Under the assumption that $x(0) = 0$, we have

$$\begin{bmatrix} y(0)\\ y(1)\\ y(2)\\ \vdots\\ y(s)\end{bmatrix} = Q\begin{bmatrix} w(0)\\ w(1)\\ w(2)\\ \vdots\\ w(s)\end{bmatrix}, \tag{2.8}$$

where

$$\begin{aligned}
Q &= \begin{bmatrix} D_w & 0 & \cdots & 0\\ CB_w & D_w & & \\ CAB_w & CB_w\ D_w & \ddots & \\ \vdots & & & \\ CA^{s-1}B_w & \cdots & CAB_w\ CB_w & D_w \end{bmatrix}\\
&= [q_{ij}] \in \mathbb{R}^{(s+1)r\times(s+1)m}.
\end{aligned}$$

Taking the 1-norm of both sides of (2.8) yields $\sum_{k=0}^{s} \|y(k)\|_1 \le \|Q\|_1 \sum_{k=0}^{s} \|w(k)\|_1$ with

$$\|Q\|_1 = \max_j \left(\sum_{i=1}^{(s+1)r} q_{ij} \right),$$

that is,

$$\|Q\|_1 = \|CB_w + CAB_w + \cdots + CA^{s-1}B_w + D_w\|_1.$$

Since A is asymptotically stable, as $s \to \infty$, $\|Q\|_1 \to \|C(I-A)^{-1}B_w + D_w\|_1$, which leads to $\|y\|_{\ell_1} \le \|C(I-A)^{-1}B_w + D_w\|_1 \|w\|_{\ell_1}$.

In the following, we investigate the condition under which $\frac{\|y\|_{\ell_1}}{\|w\|_{\ell_1}}$ reaches its supreme value. Suppose all the components of vector $w(k)$ are equal to zero except at time $k = 0$. We have

$$\begin{aligned} y(0) &= D_w w(0), \\ y(1) &= CB_w w(0), \\ y(2) &= CAB_w w(0), \\ &\vdots \\ y(s) &= CA^{s-1}B_w w(0), \end{aligned} \tag{2.9}$$

which yields $\sum_{k=0}^{s} \|y(k)\|_1 = \mathbf{1}^T(CB_w + CAB_w + \cdots + CA^{s-1}B_w + D_w)w(0)$.

Since A is asymptotically stable, as $s \to \infty$,

$$\|y\|_{\ell_1} = \mathbf{1}^T(C(I-A)^{-1}B_w + D_w)w(0) \tag{2.10}$$

$$\le \|C(I-A)^{-1}B_w + D_w\|_1 \|w\|_{\ell_1}. \tag{2.11}$$

Now denote $\bar{Q} = [\bar{q}_j] \triangleq \mathbf{1}^T(C(I-A)^{-1}B_w + D_w)$. Without loss of generality, we assume that the norm $\|\bar{Q}\|_1 = \max_j(\bar{q}_j)$ achieves its supreme value at the j^*_{th} column and all the components of vector $w(0)$ are equal to zero except at the j^*_{th} row, we have that $\frac{\|y\|_{\ell_1}}{\|w\|_{\ell_1}}$ achieves its supremal value $\|C(I-A)^{-1}B_w + D_w\|_1$. □

2.2.2 *Novel ℓ_1-Induced Performance Characterization*

In this subsection, our objective is to develop a novel characterization under which system (2.6) is asymptotically stable and satisfies the performance in (2.4).

When the input w is taken into account, we can derive the following result which provides a fundamental characterization on the stability of system (2.6) with the performance in (2.4).

Theorem 2.2 *The positive linear system in (2.6) is asymptotically stable and satisfies* $\|y\|_{\ell_1} < \gamma \|w\|_{\ell_1}$ *if and only if there exists a vector* $p \geq\geq 0$ *satisfying*

$$\mathbf{1}^T C + p^T A - p^T \ll 0, \tag{2.12}$$
$$p^T B + \mathbf{1}^T D - \gamma \mathbf{1}^T \ll 0, \tag{2.13}$$

Proof Sufficiency: First, we assume that there exists an integer k such that $x(k) \neq 0$. From (2.12), we can see that $p^T A - p^T \ll 0$ holds, and thus the asymptotic stability of system (2.6) is proved.

Consider the linear Lyapunov function candidate $V(x) = p^T x$, computing the Lyapunov difference yields

$$\Delta V(k) = p^T (Ax(k) + Bw(k)) - p^T x(k).$$

Let

$$\begin{aligned} J &= \|y(k)\|_1 - \gamma \|w(k)\|_1 \\ &= \sum_{i=1}^{r} y_i(k) - \gamma \sum_{i=1}^{m} w_i(k) \\ &= \left[\sum_{i=1}^{r} y_i(k) - \gamma \sum_{i=1}^{m} w_i(k) + \Delta V(k) \right] - \Delta V(k) \\ &= \left[\mathbf{1}^T Cx(k) + \mathbf{1}^T Dw(k) - \gamma \mathbf{1}^T w(k) \right. \\ &\quad \left. + p^T (Ax(k) + Bw(k)) - p^T x(k) \right] - \Delta V(k) \\ &= \left[\mathbf{1}^T C + p^T (A - I) \right] x(k) \\ &\quad + (\mathbf{1}^T D + p^T B - \gamma \mathbf{1}^T) w(k) - \Delta V(k) \\ &= \left[\mathbf{1}^T C + p^T (A - I) + \varepsilon \mathbf{1}^T \right] x(k) + (\mathbf{1}^T D \\ &\quad + p^T B - \gamma \mathbf{1}^T) w(k) - \varepsilon \mathbf{1}^T x(k) - \Delta V(k), \end{aligned} \tag{2.14}$$

where $\varepsilon > 0$ is sufficiently small.

From (2.12) and (2.13), we have

$$\sum_{k=0}^{s} \sum_{i=1}^{r} y_i(k) + \varepsilon \sum_{k=0}^{s} \sum_{i=1}^{n} x_i(k) < \gamma \sum_{k=0}^{s} \sum_{i=1}^{m} w_i(k) - V(s+1). \tag{2.15}$$

Since the system is asymptotically stable, we have

$$\sum_{k=0}^{\infty} \sum_{i=1}^{r} y_i(k) + \varepsilon \sum_{k=0}^{\infty} \sum_{i=1}^{n} x_i(k) \leq \gamma \sum_{k=0}^{\infty} \sum_{i=1}^{m} w_i(k), \tag{2.16}$$

which implies

$$\|y\|_{\ell_1} < \gamma\|w\|_{\ell_1}. \tag{2.17}$$

Next we consider the case with $x(k) = 0$. From (2.12), the asymptotic stability of system (2.6) is proved. It is easy to see that if $x(k) = 0$, we have $y(k) = Dw(k)$ and from (2.13), $\|y\|_{\ell_1} < \gamma\|w\|_{\ell_1}$ holds. This proves sufficiency.

Necessity: Assume that system (2.6) is asymptotically stable and satisfies the performance $\|y\|_{\ell_1} < \gamma\|w\|_{\ell_1}$. Now it follows that (2.16) holds, that is, under zero initial conditions,

$$\sum_{k=0}^{\infty}\left(\sum_{i=1}^{r} y_i(k) - \gamma\sum_{i=1}^{m} w_i(k)\right) < 0, \tag{2.18}$$

which is equal to

$$(\mathbf{1}^T D + \mathbf{1}^T C(I - A)^{-1}B - \gamma\mathbf{1}^T)\sum_{k=0}^{\infty} w(k) < 0. \tag{2.19}$$

As $0 \neq w \in \ell_1$ is arbitrary, inequality (2.19) implies

$$\mathbf{1}^T D + \mathbf{1}^T C(I - A)^{-1}B - \gamma\mathbf{1}^T \ll 0. \tag{2.20}$$

Define $\tilde{p} \triangleq (\mathbf{1}^T C(I - A)^{-1})^T \geq\geq 0$ and $p \triangleq \tilde{p} + \epsilon\alpha \gg 0$, where $\alpha \gg 0$ satisfies $\alpha^T(I - A) \gg 0$, and $\epsilon > 0$ is sufficiently small. We have

$$\begin{aligned}
\mathbf{1}^T C + p^T A - p^T &= \mathbf{1}^T C - (\tilde{p}^T + \epsilon\alpha^T)(I - A) \\
&= \mathbf{1}^T C - \mathbf{1}^T C - \epsilon\alpha^T(I - A) \\
&= -\epsilon\alpha^T(I - A) \\
&\ll 0.
\end{aligned} \tag{2.21}$$

On the other hand,

$$\mathbf{1}^T D + p^T B - \gamma\mathbf{1}^T = \mathbf{1}^T D + \mathbf{1}^T C(I - A)^{-1}B + \epsilon\alpha^T B - \gamma\mathbf{1}^T. \tag{2.22}$$

From (2.20) and ϵ is sufficiently small, we have that (2.13) holds. □

Remark 2.1 The condition obtained in Theorem 2.2 is necessary and sufficient in terms of linear programming which can be verified and solved efficiently. Hence, we can easily obtain γ and feasible solution p with the help of convex optimization techniques.

Remark 2.2 The result developed in Theorem 2.2 has also appeared in [1] and been proven by virtue of dissipativity theory. Theorem 2.2 gives an alternative proof and the result was obtained independently of the paper [1].

2.3 Controller Synthesis and Algorithm

In this section, the synthesis results of state-feedback controllers for positive system is derived with linear Lyapunov functions. Based on the analysis in Sect. 2.2, a necessary and sufficient condition for the existence of a solution to Problem PPL1CD is obtained. Then, an iterative convex optimization approach is developed to compute the controller matrices accordingly.

Corollary 2.1 *The closed-loop system (2.3) is positive, asymptotically stable and satisfies* $\|y\|_{\ell_1} < \gamma \|w\|_{\ell_1}$ *if and only if there exist a matrix* K *and a vector* $p \gg 0$ *satisfying*

$$\begin{aligned} A + BK &\geq\geq 0, && (2.23)\\ C + DK &\geq\geq 0, && (2.24)\\ \mathbf{1}^T(C + DK) + p^T(A + BK) - p^T &\ll 0, && (2.25)\\ p^T B_w + \mathbf{1}^T D_w - \gamma \mathbf{1}^T &\ll 0. && (2.26) \end{aligned}$$

Remark 2.3 Note that conditions (2.25) and (2.26) are LMIs and have the polynomial time complexity [2]. However, in (2.25), the Lyapunov vector p is coupled with the controller matrix K. It should be noted that such a problem is generally a bilinear matrix inequality (BMI) problem and known to be NP-hard, which means that it is rather unlikely to find a polynomial time algorithm for solving general BMI problems [3].

In the following, our aim is to derive a numerically tractable mean to synthesize the required controllers with the help of convex optimization. It is noted that when matrix K is fixed, (2.25) turns out to be linear with respect to the other variables. Therefore, a natural way is to fix K, and solve (2.23)–(2.26) by linear programming. Thus, the following iterative algorithm can be proposed to solve the problem (see [4]).

Algorithm PPL1CD

- Step 1. Set $i = 1$. Select an initial matrix K_1 such that system

  $$\begin{cases} x(k+1) = Ax(k) + Bu(k) + B_w w(k), \\ \quad\; y(k) = Cx(k) + Du(k) + D_w w(k), \end{cases}$$

  with

  $$u(k) = K_1 x(k) \qquad (2.27)$$

  is asymptotically stable and $A + BK_1 \geq\geq 0$ and $C + DK_1 \geq\geq 0$.
- Step 2. For fixed K_i, solve the following optimization problem for p_i and γ_i. OP2: Minimize γ_i subject to the following constraints:

$$\begin{aligned}\mathbf{1}^T(C+DK)+p_i^T(A+BK_i)-p_i^T &\ll 0,\\ p_i^TB_w+\mathbf{1}^TD_w-\gamma_i\mathbf{1}^T &\ll 0,\\ p_i &\geq\geq 0.\end{aligned}$$

Denote γ_i^*, p_i as the solution to the optimization problem. If $\left|\left(\gamma_i^*-\gamma_{i-1}^*\right)/\gamma_i^*\right| \leq \varepsilon_1$, where ε_1 is a prescribed bound, then $K=K_i$, $p=p_i$. STOP.

- Step 3. For fixed p_i, solve the following optimization problem for K_i. OP2: Minimize γ_i subject to the following constraints:

$$\begin{aligned}A+BK_i &\geq\geq 0,\\ C+DK_i &\geq\geq 0,\\ \mathbf{1}^T(C+DK)+p_i^T(A+BK_i)-p_i^T &\ll 0.\end{aligned}$$

Denote γ_i^* as the solution to the optimization problem.

- Step 4. If $\left|\left(\gamma_i^*-\gamma_{i-1}^*\right)/\gamma_i^*\right| \leq \varepsilon_2$, where ε_2 is a prescribed tolerance, STOP; else set $i=i+1$ and $K_i=K_{i-1}$, then go to Step 2.

Remark 2.4 The selection of K_1 in Step 1 can be made easily. In fact, from [5], we know that system (2.1) with (2.27) is positive and asymptotically stable if and only if there exist a diagonal matrix $P \triangleq \text{diag}(p_1, p_2, \ldots, p_n)$ and $Q \triangleq [q_{ij}] \in \mathbb{R}^{l\times n}$ such that

$$\begin{bmatrix}-P & AP+BQ\\ * & -P\end{bmatrix} < 0. \tag{2.28}$$

$$a_{ij}p_j+\sum_{z=1}^{l} b_{iz}q_{zj} \geq 0. \tag{2.29}$$

Under this condition, an initial choice of K can be given by $K_1 = QP^{-1}$.

Remark 2.5 The parameter γ can be optimized iteratively. Notice that $\gamma_{i+1}^* \leq \gamma_i^*$ since the corresponding parameters obtained in Step 3 will be utilized as the initial values in Step 2 to derive a smaller γ. Therefore, the convergence of the iterative process is naturally guaranteed. Moreover, it follows from Step 1 that if one cannot find such a matrix K_1, then it can be concluded immediately that there does not exist a solution to Problem PPL1CD.

Remark 2.6 Note that problem in Step 1 is an LMI problem, which has a polynomial time complexity [6]. For LMI problems, the number $\mathcal{N}(\varepsilon)$ of flops needed to compute an ε-accurate solution is bounded by $O(\mathcal{M}\mathcal{N}^3 log(\mathcal{V}/\varepsilon))$, where $\mathcal{M}$ is the total row size of the LMI system, $\mathcal{N}$ is the total number of scalar decision variables, $\mathcal{V}$ is a data-dependent scaling factor, and ε is the relative accuracy set for the algorithm. Every problem in Steps 2–3 is an LP problem and it is less computational demanding than LMI problems.

2.4 Analytical Method for Special Case

This section is devoted to the synthesis of the state-feedback controller for SIMO systems. It is noted that in the general case, the result in Corollary 2.1 is based on the introduction of a nonnegative vector p. Different from the synthesis in Sect. 2.3, a novel design approach is given, and some links to the convergence rate of the closed-loop systems are provided. Moreover, we extend the results to the multiple-input multiple-output (MIMO) systems by deriving the minimal K under a structural constraint, which may provide an upper bound of the ℓ_1-induced performance when the closed-loop system is stable.

First, we give the following result which is valid for general MIMO systems.

Theorem 2.3 *The closed-loop system (2.3) is positive, asymptotically stable and satisfies* $\|y\|_{\ell_1} < \gamma \|w\|_{\ell_1}$ *if and only if there exists a matrix K satisfying*

$$A + BK \geq\geq 0, \tag{2.30}$$

$$C + DK \geq\geq 0, \tag{2.31}$$

$$\rho(A + BK) < 1, \tag{2.32}$$

$$\|(C + DK)(I - (A + BK))^{-1} B_w + D_w\|_1 < \gamma. \tag{2.33}$$

Proof From (2.30)–(2.32), it is easy to see that closed-loop system (2.3) is positive and asymptotically stable.

According to Theorem 2.1, the ℓ_1-induced norm of system (2.3) is

$$\|\bar{G}\|_{\ell_1 \to \ell_1} = \|(C + DK)(I - (A + BK))^{-1} B_w + D_w\|_1. \tag{2.34}$$

From (2.33), we have $\|y\|_{\ell_1} < \gamma \|w\|_{\ell_1}$ holds.

The necessity can be obtained by reversing the above procedure and the details are omitted here. □

In the following, we present an analytic solution to the controller design problem. When (2.32) holds, we have that

$$(I - (A + BK))^{-1} = \sum_{i=0}^{\infty} (A + BK)^i. \tag{2.35}$$

If the elements of $A + BK \geq\geq 0$ are sufficiently small, (2.32) is more likely to be satisfied and from (2.34), the ℓ_1-induced norm of system (2.3) will also be small when all the elements of $A + BK \geq\geq 0$ and $C + DK \geq\geq 0$ are sufficiently small. For SIMO systems, the problem can be analytically solved.

For ease of reference when addressing SIMO systems, we have B and D as vectors and write

$$B = b = \begin{bmatrix} b_1 \\ b_2 \\ \vdots \\ b_n \end{bmatrix}, \quad D = d = \begin{bmatrix} d_1 \\ d_2 \\ \vdots \\ d_r \end{bmatrix},$$

where $b \in \bar{\mathbb{R}}^n_+$ and $d \in \bar{\mathbb{R}}^r_+$. In this case, K is a row vector given by

$$K = \begin{bmatrix} k_1 \; k_2 \; \dots \; k_n \end{bmatrix}.$$

We will show in the following theorem that for SIMO systems, the optimal K can be computed analytically. Moreover, if such an optimal K gives an asymptotically stable closed-loop system, it achieves the minimal performance level γ^* and the closed-loop system has the fastest convergence rate.

Theorem 2.4 *For an SIMO system with $b \neq 0$, the optimal K such that the elements of $A + bK \geq\geq 0$ and $C + dK \geq\geq 0$ are jointly minimized is given by*

$$K^* = \begin{bmatrix} k_1^* \; k_2^* \; \dots \; k_n^* \end{bmatrix} \tag{2.36}$$

where

$$k_j^* = - \min_{\substack{i=1,2,\dots,n \\ h=1,2,\dots,r}} \left\{ \frac{a_{ij}}{b_i}, \frac{c_{hj}}{d_h} \right\} \tag{2.37}$$

with the corresponding term inside $\{\cdot\}$ disappears if $b_i = 0$ or $d_h = 0$. Moreover, if $\rho(A + bK^*) < 1$, then $K^* = \arg\min \|(C + dK)(I - (A + bK))^{-1} B_w + D_w\|_1$ subject to $A + bK \geq\geq 0$ and $C + dK \geq\geq 0$ and the closed-loop system has the fastest convergence rate.

Proof Observe that the constraints $A+bK \geq\geq 0$ and $C+dK \geq\geq 0$ on k_j come from the jth column of the two matrices only. With $A = [a_{ij}] \geq\geq 0$ and $C = [c_{hj}] \geq\geq 0$, these constraints on k_j can be written explicitly as

$$\begin{aligned} a_{ij} + b_i k_j &\geq 0, \qquad i = 1, 2, \dots, n, \\ c_{hj} + d_h k_j &\geq 0, \qquad h = 1, 2, \dots, r. \end{aligned}$$

It is easy to see that if $b_i = 0$ or $d_h = 0$, the corresponding constraint is slack. In fact, the remaining constraints become

$$k_j \geq \max_{\substack{i=1,2,\dots,n \\ h=1,2,\dots,r}} \left\{ -\frac{a_{ij}}{b_i}, -\frac{c_{hj}}{d_h} \right\} = - \min_{\substack{i=1,2,\dots,n \\ h=1,2,\dots,r}} \left\{ \frac{a_{ij}}{b_i}, \frac{c_{hj}}{d_h} \right\}. \tag{2.38}$$

Thus, to minimize the elements of $A + bK$ and $C + dK$, the choice of k_j should be the minimum value subject to constraints in (2.38) which is given by k_j^* in (2.37). Hence, the first part of the result has been established.

When $\rho(A + bK^*) < 1$, the system is asymptotically stable and

$$(I - (A + bK^*))^{-1} = \sum_{i=0}^{\infty} (A + bK^*)^i.$$

For any $K \geq\geq K^*$ satisfying $\rho(A + bK) < 1$, we have

$$(C + dK)(I - (A + bK))^{-1} B_w + D_w = \sum_{i=0}^{\infty} (C + dK)(A + bK)^i B_w + D_w.$$

Since $0 \leq\leq A + bK^* \leq\leq A + bK$ and $0 \leq\leq C + dK^* \leq\leq C + dK$, we have, from Lemma 1.6, that

$$0 \leq\leq (C + dK^*)(A + dK^*)^i \leq\leq (C + dK)(A + dK)^i$$

and consequently,

$$\begin{aligned} 0 &\leq\leq (C + dK^*)(I - (A + bK^*))^{-1} B_w + D_w \\ &\leq\leq (C + dK)(I - (A + bK))^{-1} B_w + D_w. \end{aligned}$$

The 1-norm of these matrices thus gives

$$\begin{aligned} &\|(C + dK^*)(I - (A + bK^*))^{-1} B_w + D_w\|_1 \\ &\leq \|(C + dK)(I - (A + bK))^{-1} B_w + D_w\|_1. \end{aligned}$$

As K^* gives the smallest K elementwise satisfying the constraints $A+bK \geq\geq 0$ and $C + dK \geq\geq 0$, the value of the ℓ_1-induced performance corresponds to its minimal value.

Finally, from Lemma 1.3, we have $\rho(A+bK^*) \leq \rho(A+bK)$ for any $A+bK \geq\geq 0$. The convergence rate is characterized by the spectral radius. Hence, the closed-loop system with K^* has the fastest convergence rate. The proof is completed. □

Remark 2.7 For SIMO systems with $b \neq 0$, Theorem 2.4 gives a complete solution for the minimal ℓ_1-induced performance problem when $\rho(A + bK^*) < 1$. Notice that the optimal K in (2.36) gives the fastest convergence rate when the closed-loop system is stable, but it does not minimize the spectral radius of $A + bK$ due to the presence of the constraint $C + dK \geq\geq 0$.

It can be easily seen that if $d = 0$, then the constraint $C + dK \geq\geq 0$ is naturally satisfied and

$$k_j^* = - \min_{i=1,2,\ldots,n} \left\{ \frac{a_{ij}}{b_i} \right\} \tag{2.39}$$

with the corresponding term inside $\{\cdot\}$ disappears if $b_i = 0$.

In this case, we have the following corollary in view of Remark 2.7.

Corollary 2.2 *For an SIMO system with $b \neq 0$ and $d = 0$, the optimal K given by (2.36) with k_j^* in (2.39) minimizes $\rho(A + bK)$ subject to $A + bK \geq\geq 0$. If $\rho(A + bK^*) < 1$, then the closed-loop system achieves the minimal ℓ_1-induced performance.*

We note that actually Theorem 2.3 serves as a characterization of the existence of a state-feedback controller for general MIMO systems. Unfortunately, for multi-input systems, the problem of jointly minimizing all the elements of $A + BK$ and $C + DK$ is not generally well-posed. Under a structural constraint, we present a method to derive the minimal K for MIMO systems.

Suppose $K \in \mathbb{R}^{m\times n}$ takes the following structure:

$$K = L\Delta \tag{2.40}$$

where $L \triangleq \begin{bmatrix} l_1 \ l_2 \ \dots \ l_n \end{bmatrix} \in \bar{\mathbb{R}}_+^{m\times n}$ is a fixed 'direction matrix' with $l_j \geq\geq 0$, $j = 1, 2, \dots, n$, being unit vectors, and $\Delta \triangleq \mathrm{diag}(\delta_1, \delta_2, \dots, \delta_n)$ with δ_j to be determined. For convenience, we write

$$B = \begin{bmatrix} \bar{b}_1^T \\ \bar{b}_2^T \\ \vdots \\ \bar{b}_n^T \end{bmatrix}, \quad D = \begin{bmatrix} \bar{d}_1^T \\ \bar{d}_2^T \\ \vdots \\ \bar{d}_r^T \end{bmatrix}$$

and there hold $\bar{b}_i^T l_j \geq 0$ and $\bar{d}_h^T l_j \geq 0$.

The following proposition provides a characterization of the optimal structural K which jointly minimizes the constraints $A + BK \geq\geq 0$ and $C + DK \geq\geq 0$. The proof is similar to that of Theorem 2.4 and hence omitted.

Proposition 2.1 *For an MIMO system with $Bl_j \neq 0$, the optimal structural K in (2.40) such that the elements of $A + BK \geq\geq 0$ and $C + DK \geq\geq 0$ are jointly minimized has Δ given by*

$$\Delta^* = \mathrm{diag}(\delta_1^*, \delta_2^*, \dots, \delta_n^*)$$

where

$$\delta_j^* = - \min_{\substack{i=1,2,\dots,n \\ h=1,2,\dots,r}} \left\{ \frac{a_{ij}}{\bar{b}_i^T l_j}, \frac{c_{hj}}{\bar{d}_h^T l_j} \right\}$$

with the corresponding term inside $\{\cdot\}$ disappears if $\bar{b}_i^T l_j = 0$ or $\bar{d}_h^T l_j = 0$.

Remark 2.8 It is noted that $A + BK \geq\geq 0$ and $C + DK \geq\geq 0$ are jointly minimized with a given direction matrix L. When $\rho(A + BL\Delta^*) < 1$, the ℓ_1-induced performance will be minimized under the structurally constrained K. However, if the closed-loop system is unstable, one has to choose another direction matrix L to search a possible K to stabilize the system. The selection of an appropriate direction matrix L to minimize the ℓ_1-induced performance represents a future research direction.

2.5 Numerical Example

In this section, we present an illustrative example to demonstrate the applicability of the proposed results.

Among different age-structural population models, the Leslie model is the most classical and widely used in population dynamics and control. In a Leslie model, the survival rates and fertility rates depend exclusively on age [7, 8]. From [9], we know that under some conditions, for example, there is an ample food supply, plenty of space and a freedom from predators in the area, it might be assumed for theoretical purposes that some age specific rates of fertility and mortality would remain approximately constant over a period of time.

In this example, we investigate the structural population dynamics of a certain pest described by a Leslie model. An external disturbance is brought into consideration and our aim is to annihilate the pests in a certain area. We consider the following Leslie model:

$$\begin{cases} x(k+1) = Ax(k) + Bu(k) + B_w w(k), \\ y(k) = Cx(k) + Du(k), \end{cases} \tag{2.41}$$

where

$$A = \begin{bmatrix} 0.2 & 0.3 & 2 \\ 0.8 & 0 & 0 \\ 0 & 0.7 & 0 \end{bmatrix}, \quad B = \begin{bmatrix} 0.5 \\ 0 \\ 0 \end{bmatrix},$$

$$B_w = \begin{bmatrix} 0.1 \\ 0.05 \\ 0.1 \end{bmatrix}, \quad C = \begin{bmatrix} 1 & 1 & 1 \end{bmatrix}, \quad D = 0.5.$$

In this model, $x(k) = \begin{bmatrix} x_1(k) & x_2(k) & x_3(k) \end{bmatrix}^T$ where $x_i(k)$ represents the number of individuals of age i in year k before the reproduction season. The external input, denoted as $w(k)$, is regarded as a measure of the population of the pests from other regions that migrates into the area of interest. The output $y(k)$ denotes the sum of the number of pests in the area.

Since the Leslie model in (2.41) is a single-input system, we first use the method proposed in Sect. 2.4 to compute the optimal K. By resorting to Theorem 2.4, we can easily obtain the controller gain matrix

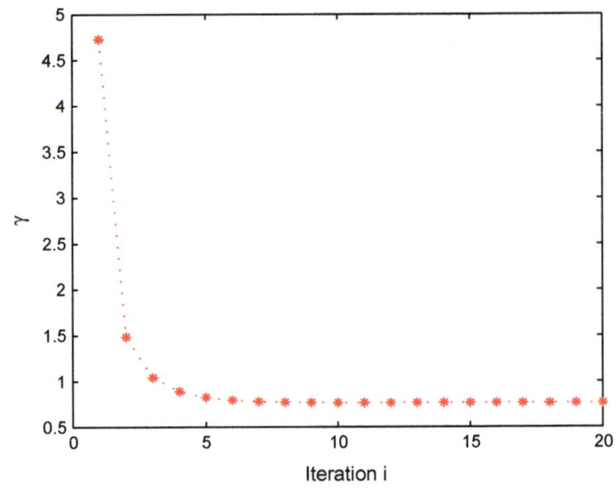

Fig. 2.1 Variation of γ_i^* with iteration i. Reprinted from Ref. [10], Copyright 2013, with permission from Elsevier

$$K = \begin{bmatrix} -0.4 & -0.6 & -2 \end{bmatrix}, \tag{2.42}$$

which yields

$$A + BK = \begin{bmatrix} 0 & 0 & 1 \\ 0.8 & 0 & 0 \\ 0 & 0.7 & 0 \end{bmatrix}, \ \rho(A + BK) = 0.8243, \ C + DK = \begin{bmatrix} 0.8 & 0.7 & 0 \end{bmatrix}.$$

With the introduction of controller (2.42), the minimal value γ^* is 0.7614 and the convergence rate of the closed-loop system is the fastest.

On the other hand, by solving conditions in Corollary 2.1 using Algorithm PPL1CD, we obtain an upper bound of γ^* as 0.7623 after 5 iterations and a feasible solution is achieved with

$$p = \begin{bmatrix} 3.0931 & 2.8658 & 3.0934 \end{bmatrix}^T,$$

which yields the controller gain matrix as

$$K = \begin{bmatrix} -0.3999 & -0.5999 & -1.9999 \end{bmatrix}.$$

Figure 2.1 shows the variation of γ_i^* with iteration i. From Fig. 2.1 one can clearly see that γ_i^* is monotonically decreasing with respect to iteration i.

The performance of the open-loop and the closed-loop system is evaluated via simulation. The initial condition used in the simulation is

$$x(0) = \begin{bmatrix} 50 & 10 & 30 \end{bmatrix}^T.$$

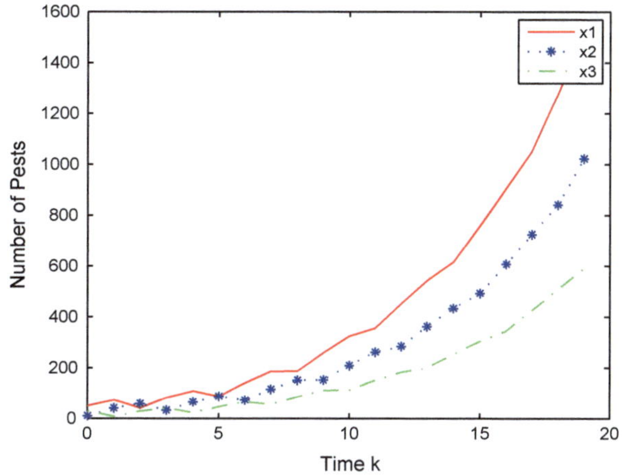

Fig. 2.2 Open-loop unforced response. Reprinted from Ref. [10], Copyright 2013, with permission from Elsevier

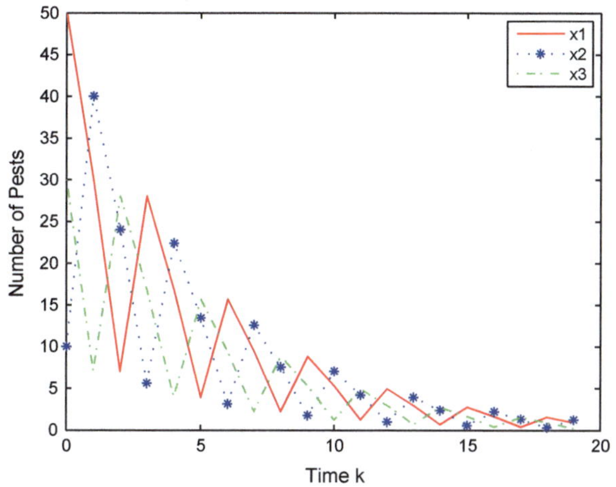

Fig. 2.3 Closed-loop unforced response. Reprinted from Ref. [10], Copyright 2013, with permission from Elsevier

Figure 2.2 shows the state response of open-loop system and Fig. 2.3 shows the state response of the closed-loop system when the external disturbance $w(k) \equiv 0$, from which we can see that the state converges to zero. To illustrate the disturbance attenuation performance, the external disturbance $w(k)$ is assumed to be

$$w(k) = \begin{cases} 50, & 5 \leq k \leq 10, \\ 0, & \text{otherwise.} \end{cases} \tag{2.43}$$

Figure 2.4 shows the response of closed-loop system with $w(k)$ and Fig. 2.5 shows the signal $w(k)$.

It is noted that for SIMO systems, we can easily compute the optimal K and the exact value of the minimal γ by resorting to Theorem 2.4. Moreover, we see that the

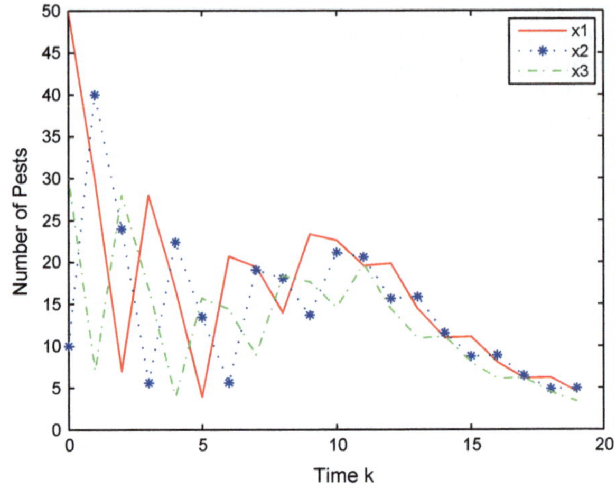

Fig. 2.4 Closed-loop forced response. Reprinted from Ref. [10], Copyright 2013, with permission from Elsevier

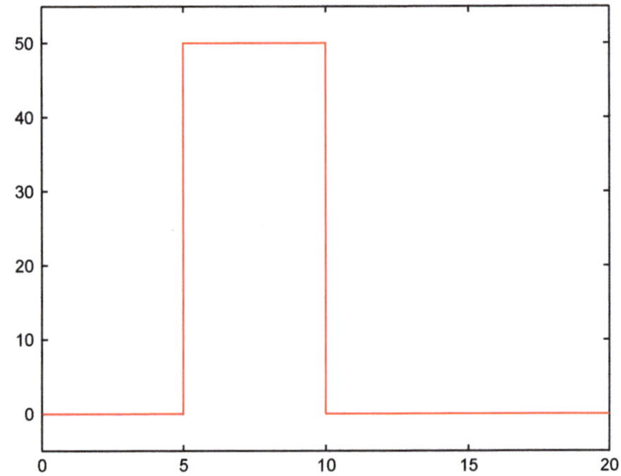

Fig. 2.5 Signal $w(k)$. Reprinted from Ref. [10], Copyright 2013, with permission from Elsevier

values of K and γ^* obtained by two approaches are very close, which implies that Corollary 2.1 also provides a good solution to the controller synthesis problem.

From this example, we know by applying the state-feedback control, even under the influx of pests from other regions, the pests in this region can be annihilated eventually. In practice, the spraying of pesticide is usually one of the pest control methods. Here, the number of insects of all ages can be reduced by spraying pesticide, which corresponds to the introduction of state-feedback controller $u(k) = Kx(k)$ where K represents the amount of pesticide.

2.6 Summary

In this chapter, the problem of state-feedback controller design for linear positive systems with the use of linear Lyapunov function has been studied. A method has been derived to compute the exact value of ℓ_1-induced norm for positive systems. A characterization has been proposed to ensure the asymptotic stability of the controlled system with a prescribed ℓ_1-induced performance level. A necessary and sufficient condition for the existence of a desired controller has been established accordingly. Then, an iterative convex optimization algorithm has been developed to solve the design conditions. In addition, an analytical design approach has been put forward to investigate the synthesis of the state-feedback controller for SIMO positive systems. Finally, one example has been presented to illustrate the effectiveness of the theoretical results.

References

1. Briat C (2013) Robust stability and stabilization of uncertain linear positive systems via integral linear constraints: L_1-gain and L_∞-gain characterization. Int J Robust Nonlinear Control 23(17):1932–1954
2. Gahinet P, Nemirovski A, Laub A et al (1995) LMI control toolbox user's guide. The mathworks, Natick
3. Toker O, Ozbay H (1995) On the NP-hardness of solving bilinear matrix inequalities and simultaneous stabilization with static output feedback. Proceedings of american control conference 4:2525–2526
4. Cao YY, Lam J, Sun YX et al (1998) Static output feedback stabilization: an ILMI approach. Automatica 34(12):1641–1645
5. Gao H, Lam J, Wang C et al (2005) Control for stability and positivity: equivalent conditions and computation. IEEE Trans Circuits Syst (II) 52(9):540–544
6. Fornasini E, Valcher ME (2010) Linear copositive Lyapunov functions for continuous-time positive switched systems. IEEE Trans. Automat. Control 55:1933–1937
7. Muratori S, Rinaldi S (1990) Equilibria, stability, and reachability of Leslie systems with nonnegative inputs. IEEE Trans Automat Control 35(9):1065–1068
8. Farina L, Rinaldi S (2011) Positive linear systems: theory and applications. Wiley, Hoboken
9. Leslie PH (1945) On the use of matrices in certain population mathematics. Biometrika 33(3):183–212
10. Chen X, Lam J, Li P et al (2013) ℓ_1-induced norm and controller synthesis of positive systems-induced norm and controller synthesis of positive systems. Automatica 49:1377–1385

Chapter 3
L_1-Induced Output-Feedback Controller Synthesis for Interval Positive Systems

This chapter is concerned with the design of L_1-induced output-feedback controller for continuous-time positive systems with interval uncertainties. A necessary and sufficient condition for stability and L_1-induced performance of positive linear systems is proposed in terms of linear inequalities. Based on this, conditions for the existence of robust static output-feedback controllers are established and an iterative convex optimization approach is developed to solve the conditions. The problem of controller synthesis is completely solved for single-input-multiple-output (SIMO) positive systems with the help of an analytical formula for L_1-induced norm. The result is further extended to the multiple-input-multiple-output (MIMO) case by employing a structured controller. Moreover, the design of L_1-induced sparse controller is investigated for continuous-time interval positive systems. In addition, a dynamic output-feedback controller is designed for interval positive systems under L_1 performance. Three illustrative examples are provided to show the effectiveness and applicability of the theoretical results.

The organization of this chapter is as follows. In Sect. 3.1, the exact value of L_1-induced norm is computed for continuous-time positive system. Then, the L_1-induced performance characterization is provided for interval positive system. In Sect. 3.2, a static output-feedback controller design method for interval positive systems is put forward based on the analysis conditions. Section 3.3 considers a special case of interval positive systems. Under the special case of SIMO interval positive systems, an analytical method for controller synthesis is proposed. Then, the results are further extended to the MIMO case. Moreover, the L_1-induced sparse controller is designed for continuous-time positive systems with interval uncertainties. In Sect. 3.4, we turn to consider the dynamic output-feedback stabilization problem for interval positive systems. Section 3.5 presents several illustrative examples. Section 3.6 concludes the chapter.

X. Chen, *Analysis and Synthesis of Positive Systems Under ℓ_1 and L_1 Performance*, Springer Theses, DOI 10.1007/978-981-10-2227-2_3

3.1 Performance Analysis

Consider a positive system:

$$\begin{cases} \dot{x}(t) = Ax(t) + B_w w(t), \\ y(t) = Cx(t) + D_w w(t), \end{cases} \tag{3.1}$$

where $x(t) \in \mathbb{R}^n$, $w(t) \in \mathbb{R}^m$ and $y(t) \in \mathbb{R}^r$ are the system state, input and output, respectively.

Now, we are in a position to give the definition of L_1-induced norm. For a stable positive linear system given in (3.1), its L_1-induced norm is defined as

$$\|\Im\|_{(L_1,L_1)} \triangleq \sup_{w\neq 0,\ w\in L_1} \frac{\|y\|_{L_1}}{\|w\|_{L_1}}, \tag{3.2}$$

where $\Im : L_1 \to L_1$ denotes the convolution operator, that is, $y(t) = (\Im * w)(t)$. System (3.1) is said to have L_1-induced performance at the level γ if, under zero initial conditions,

$$\|\Im\|_{(L_1,L_1)} < \gamma, \tag{3.3}$$

where $\gamma > 0$ is a given scalar.

First, we give the following theorem through which the value of L_1-induced norm of system (3.1) can be computed directly.

Theorem 3.1 *For a stable positive linear system given in (3.1), the exact value of the L_1-induced norm is given by*

$$\|\Im\|_{(L_1,L_1)} = \|D_w - CA^{-1}B_w\|_1. \tag{3.4}$$

Proof For system (3.1), the impulse response $G(t)$ is given by

$$G(t) \triangleq \begin{cases} 0,\ t < 0, \\ Ce^{At}B_w + D_w\delta(t),\ t \geq 0. \end{cases} \tag{3.5}$$

Next, let $\Im : L_1 \to L_1$ denote the convolution operator:

$$z(t) = (\Im * w)(t) \triangleq \int_0^{\infty} G(t-\tau)w(\tau)d\tau. \tag{3.6}$$

From [1], we have

$$\|\Im\|_{(L_1,L_1)} = \max_{j=1,\ldots,m} \|\mathrm{col}_j(\bar{G})\|_1, \tag{3.7}$$

where

$$\bar{G}_{ij} = \int_0^{\infty} G_{ij}(t)dt. \tag{3.8}$$

Let c_i, b_{wj}, d_{wij} denote the ith row vector, the jth column vector and the (i, j)-element of matrices C, B_w and D_w, respectively. Equation (3.8) can be written as

$$\bar{G}_{ij} = \int_0^{\infty} \left(c_i e^{At} b_{wj} + d_{wij}\delta(t)\right) dt = -c_i A^{-1} b_{wj} + d_{wij}, \tag{3.9}$$

yielding (3.4). □

Then, our objective is to develop a novel characterization under which system (3.1) is asymptotically stable and satisfies the performance in (3.3).

Theorem 3.2 *The positive linear system in (3.1) is asymptotically stable and satisfies* $\|y\|_{L_1} < \gamma\|w\|_{L_1}$ *if and only if there exists a vector* $p \geq\geq 0$ *satisfying*

$$\mathbf{1}^T C + p^T A \ll 0, \tag{3.10}$$

$$p^T B_w + \mathbf{1}^T D_w - \gamma \mathbf{1}^T \ll 0. \tag{3.11}$$

Proof (*Sufficiency*) In the following, we consider two cases: $x(t) \equiv 0$ and there exists a t such that $x(t) \neq 0$. First, for $x(t) \equiv 0$, from (3.10), the asymptotic stability of system (3.1) is proved. It is easy to see that if $x(t) \equiv 0$, we have $y(t) = D_w w(t)$ and from (3.11), $\|y\|_{L_1} < \gamma\|w\|_{L_1}$ holds.

Next, we assume that there exists a t such that $x(t) \neq 0$. From (3.10), the asymptotic stability of system (3.1) is proved.

Consider the linear Lyapunov function candidate $V(x(t)) = p^T x(t)$ and we have

$$\frac{dV(x(t))}{dt} = p^T(Ax(t) + B_w w(t)).$$

Let

$$\begin{aligned}
J &= \int_0^T \|y(t)\|_1 dt - \int_0^T \gamma\|w(t)\|_1 dt \\
&= \int_0^T \left(\sum_{i=1}^{r} y_i(t) - \gamma\sum_{i=1}^{m} w_i(t)\right) dt \\
&= \int_0^T \left(\sum_{i=1}^{r} y_i(t) - \gamma\sum_{i=1}^{m} w_i(t) + \frac{dV(t)}{dt}\right) dt - V(T) \\
&= \int_0^T \left(\left[\mathbf{1}^T C + p^T A\right] x(t) + (\mathbf{1}^T D_w + p^T B_w - \gamma\mathbf{1}^T) w(t)\right) dt - V(T)
\end{aligned}$$

$$
\begin{aligned}
&= \int_0^T \left(\left[\mathbf{1}^T C + p^T A + \varepsilon \mathbf{1}^T\right] x(t) + (\mathbf{1}^T D_w + p^T B_w - \gamma \mathbf{1}^T) w(t) \right) dt \\
&\quad - \int_0^T \varepsilon \mathbf{1}^T x(t) dt - V(T),
\end{aligned}
$$

where $\varepsilon > 0$ is sufficiently small such that $\mathbf{1}^T C + p^T A + \varepsilon \mathbf{1}^T \ll 0$ holds.

From (3.10) and (3.11), we have

$$
J + \int_0^T \varepsilon \mathbf{1}^T x(t) dt + V(T) < 0,
$$

which equals to

$$
\int_0^T \left(\sum_{i=1}^r y_i(t) \right) dt + \varepsilon \int_0^T \left(\sum_{i=1}^n x_i(t) \right) dt < \gamma \int_0^T \left(\sum_{i=1}^m w_i(t) \right) dt - p^T x(T).
$$

Since the system is asymptotically stable, when $T \to \infty$ we have

$$
\int_0^\infty \left(\sum_{i=1}^r y_i(t) \right) dt + \varepsilon \int_0^\infty \left(\sum_{i=1}^n x_i(t) \right) dt \leq \gamma \int_0^\infty \left(\sum_{i=1}^m w_i(t) \right) dt,
$$

which implies

$$
\|y\|_{L_1} < \gamma \|w\|_{L_1}.
$$

This proves sufficiency.

(*Necessity*) Assume that system (3.1) is asymptotically stable and satisfies $\|y\|_{L_1} < \gamma \|w\|_{L_1}$. Then, according to Theorem 3.1, the following inequality holds

$$
\|D_w - C A^{-1} B_w\|_1 < \gamma, \tag{3.12}
$$

which implies

$$
\mathbf{1}^T D_w - \mathbf{1}^T C A^{-1} B_w - \gamma \mathbf{1}^T \ll 0. \tag{3.13}
$$

Define $\tilde{p} \triangleq (-\mathbf{1}^T C A^{-1})^T \geq\geq 0$ and $p \triangleq \tilde{p} + \epsilon \xi \gg 0$, where $\xi \gg 0$ satisfies $\xi^T(-A) \gg 0$, and $\epsilon > 0$ is a sufficiently small number. We have

$$
\begin{aligned}
\mathbf{1}^T C + p^T A &= \mathbf{1}^T C + (\tilde{p}^T + \epsilon \xi^T) A \\
&= \mathbf{1}^T C - \mathbf{1}^T C + \epsilon \xi^T A \\
&= \epsilon \xi^T A \\
&\ll 0.
\end{aligned} \tag{3.14}
$$

On the other hand,

$$\mathbf{1}^T D_w + p^T B_w - \gamma \mathbf{1}^T = \mathbf{1}^T D_w - \mathbf{1}^T C A^{-1} B_w + \epsilon \xi^T B_w - \gamma \mathbf{1}^T.$$

From (3.13) and $\epsilon > 0$ is sufficiently small, we have that (3.11) holds. This completes the whole proof. □

Remark 3.1 Similar to Theorem 2.2 for discrete-time positive systems, the condition obtained in Theorem 3.2 is necessary and sufficient in terms of linear programming. Therefore, they can be verified and solved efficiently with the help of convex optimization techniques.

3.2 Static Output-Feedback Controller Design

This section deals with the robust static output-feedback stabilization problem for positive systems with interval uncertainties. Based on the analysis results in Sect. 3.1, a static output-feedback controller is designed for continuous-time positive systems.

In the following, we consider the positive interval system

$$\mathcal{S}_{\mathcal{I}} : \begin{cases} \dot{x}(t) = Ax(t) + Bu(t) + B_w w(t), \\ z(t) = C_z x(t) + D_z u(t) + D_{zw} w(t), \\ y(t) = Cx(t), \end{cases} \tag{3.15}$$

where the system matrices A, B_w, C_z, D_{zw} and C are not precisely known, but belong to the following interval uncertainty domain:

$$A \in [\underline{A}, \overline{A}],\ B_w \in [\underline{B}_w, \overline{B}_w],\ C_z \in [\underline{C}_z, \overline{C}_z],\ D_{zw} \in [\underline{D}_{zw}, \overline{D}_{zw}],\ C \in [\underline{C}, \overline{C}]. \tag{3.16}$$

System (3.15) is said to be positive and robustly stable if it is positive and asymptotically stable over all admissible uncertainty domain in (3.16).

Here, we provide a theorem as the performance characterization for positive system $\mathcal{S}_{\mathcal{I}}$ over the whole uncertain domain in (3.16).

Theorem 3.3 *Suppose $\mathcal{S}_{\mathcal{I}}$ is positive. $\mathcal{S}_{\mathcal{I}}$ is robustly stable and satisfies $\|z\|_{L_1} < \gamma \|w\|_{L_1}$ for any $A \in [\underline{A}, \overline{A}]$, $C_z \in [\underline{C}_z, \overline{C}_z]$, $B_w \in [\underline{B}_w, \overline{B}_w]$ and $D_{zw} \in [\underline{D}_{zw}, \overline{D}_{zw}]$ under $x(0) = 0$ if and only if there exists a vector $p \geq\geq 0$ satisfying*

$$\mathbf{1}^T \overline{C}_z + p^T \overline{A} \ll 0, \tag{3.17}$$

$$p^T \overline{B}_w + \mathbf{1}^T \overline{D}_{zw} - \gamma \mathbf{1}^T \ll 0. \tag{3.18}$$

Proof (*Sufficiency*) For any $A \in [\underline{A}, \overline{A}]$, $C_z \in [\underline{C}_z, \overline{C}_z]$, $B_w \in [\underline{B}_w, \overline{B}_w]$ and $D_{zw} \in [\underline{D}_{zw}, \overline{D}_{zw}]$,

$$\begin{aligned} \mathbf{1}^T C_z + p^T A &\leq\leq \mathbf{1}^T \overline{C}_z + p^T \overline{A} \ll 0, \\ p^T B_w + \mathbf{1}^T D_{zw} - \gamma \mathbf{1}^T &\leq\leq p^T \overline{B}_w + \mathbf{1}^T \overline{D}_{zw} - \gamma \mathbf{1}^T \ll 0, \end{aligned} \tag{3.19}$$

which, by Theorem 3.2, implies that $\mathcal{S}_{\mathcal{I}}$ is robustly stable and satisfies $\|z\|_{L_1} < \gamma\|w\|_{L_1}$ over all interval uncertainty domain under $x(0) = 0$. This proves sufficiency.

(*Necessity*) Assume that $\mathcal{S}_{\mathcal{I}}$ is robustly stable and satisfies $\|z\|_{L_1} < \gamma\|w\|_{L_1}$ under $x(0) = 0$. From Theorem 3.2, we have

$$\begin{aligned} \mathbf{1}^T C_z + p^T A &\ll 0, \\ p^T B_w + \mathbf{1}^T D_{zw} - \gamma \mathbf{1}^T &\ll 0, \end{aligned}$$

which implies that (3.17) and (3.18) hold. This completes the whole proof. □

In the following, the problem of L_1-induced static output-feedback controller design (L1SOFCD) is formulated.

Problem L1SOFCD: *Given $\mathcal{S}_{\mathcal{I}}$ positive. Find a static output-feedback controller $u(t) = Ky(t)$ such that the closed-loop system*

$$\mathcal{S}_C : \begin{cases} \dot{x}(t) = (A + BKC)x(t) + B_w w(t), \\ z(t) = (C_z + D_z KC)x(t) + D_{zw} w(t), \end{cases} \tag{3.20}$$

is positive, robustly stable, and satisfies the L_1-induced performance in (3.3) *under* $x(0) = 0$.

Based on the analysis results in Theorem 3.3, we present a condition for the existence of a solution to Problem L1SOFCD.

Theorem 3.4 *Given $\mathcal{S}_{\mathcal{I}}$ positive and suppose $K \preceq\preceq 0$. $\mathcal{S}_C$ is positive, robustly stable and satisfies $\|z\|_{L_1} < \gamma\|w\|_{L_1}$ for any $A \in [\underline{A}, \overline{A}]$, $B \in [\underline{B}, \overline{B}]$, $B_w \in [\underline{B}_w, \overline{B}_w]$, $C_z \in [\underline{C}_z, \overline{C}_z]$, $D_z \in [\underline{D}_z, \overline{D}_z]$, $D_{zw} \in [\underline{D}_{zw}, \overline{D}_{zw}]$ and $C \in [\underline{C}, \overline{C}]$ under $x(0) = 0$ if and only if there exist a matrix K and a vector $p \succeq\succeq 0$ satisfying*

$$\underline{A} + \overline{B}K\overline{C} \text{ is Metzler,} \tag{3.21}$$

$$\underline{C}_z + \overline{D}_z K\overline{C} \succeq\succeq 0, \tag{3.22}$$

$$\mathbf{1}^T(\overline{C}_z + \underline{D}_z K\underline{C}) + p^T(\overline{A} + \underline{B}K\underline{C}) \ll 0, \tag{3.23}$$

$$p^T\overline{B}_w + \mathbf{1}^T\overline{D}_{zw} - \gamma\mathbf{1}^T \ll 0. \tag{3.24}$$

Proof Combining (3.21) and (3.22) with $K \preceq\preceq 0$ yields the following: for any $A \in [\underline{A}, \overline{A}]$, $B \in [\underline{B}, \overline{B}]$, $C_z \in [\underline{C}_z, \overline{C}_z]$, $D_z \in [\underline{D}_z, \overline{D}_z]$ and $C \in [\underline{C}, \overline{C}]$,

$$A + BKC \succeq\succeq \underline{A} + \overline{B}K\overline{C} \text{ is Metzler,} \quad C_z + D_z KC \succeq\succeq \underline{C}_z + \overline{D}_z K\overline{C} \succeq\succeq 0, \tag{3.25}$$

which shows that the closed-loop system (3.20) is a positive system.

It follows from (3.23) and (3.24), by Theorem 3.3, implies that the closed-loop system (3.20) is robustly stable and satisfies $\|z\|_{L_1} < \gamma\|w\|_{L_1}$.

The necessity can be obtained by reversing the above procedure and the proof is omitted here. □

Remark 3.2 We see that if the assumption $K \leq\leq 0$ is not satisfied, the inequalities in (3.25) do not necessarily hold, which leads to difficulty in the synthesis of the desired controller for interval positive system (3.15). The non-positiveness of K facilitates the controller synthesis problem. It is noted that for a special case when B, C and D_z are known constant matrices, the constraint $K \leq\leq 0$ can be removed. As for designing a sign-indefinite K, it remains unsolved and needs further study.

Note that the Lyapunov vector p is coupled with the controller matrix K in (3.23), which cannot be easily solved. However, when matrix K is fixed, (3.23) turns out to be linear with respect to the remaining variables. Therefore, a natural way is to fix K, and solve (3.21)–(3.24) by linear programming. Thus, the following iterative algorithm can be proposed to solve the conditions in Theorem 3.4.

Algorithm L1SOFCD

1. Set $i = 1$. Select an initial matrix $K_1 \leq\leq 0$ such that the positive system

$$\begin{cases} \dot{x}(t) = Ax(t) + Bu(t), \\ z(t) = C_z x(t) + D_z u(t), \\ y(t) = Cx(t), \end{cases} \tag{3.26}$$

with

$$u(t) = K_1 y(t) \tag{3.27}$$

is positive and robustly stable.

2. For fixed $K_i \leq\leq 0$, solve the following optimization problem for $p_i \geq\geq 0$ and $\gamma_i > 0$.
OP1: Minimize γ_i subject to the following constraints:

$$\mathbf{1}^T(\overline{C}_z + \underline{D}_z K_i \underline{C}) + p_i^T(\overline{A} + \underline{B} K_i \underline{C}) \ll 0, \tag{3.28}$$

$$p_i^T \overline{B}_w + \mathbf{1}^T \overline{D}_{zw} - \gamma_i \mathbf{1}^T \ll 0. \tag{3.29}$$

$$p_i \geq\geq 0. \tag{3.30}$$

Denote γ_i^*, p_i as the solution to the optimization problem.
If $\left|(\gamma_i^* - \gamma_{i-1}^*)/\gamma_i^*\right| < \varepsilon_1$, where ε_1 is a prescribed bound, then $K_{opt} = K_i$, $p = p_i$. STOP.

3. For fixed p_i, solve the following optimization problem for $K_i \leq\leq 0$.
OP2: Minimize γ_i subject to the following constraints:

$$\underline{A} + \overline{B} K_i \overline{C} \text{ is Metzler}, \tag{3.31}$$

$$\underline{C}_z + \overline{D}_z K_i \overline{C} \geq\geq 0, \tag{3.32}$$

$$\mathbf{1}^T(\overline{C}_z + \underline{D}_z K_i \underline{C}) + p_i^T(\overline{A} + \underline{B} K_i \underline{C}) \ll 0, \tag{3.33}$$

$$p_i^T \overline{B}_w + \mathbf{1}^T \overline{D}_{zw} - \gamma_i \mathbf{1}^T \ll 0. \tag{3.34}$$

Denote γ_i^* as the solution to the optimization problem.
If $\left|\left(\gamma_i^* - \gamma_{i-1}^*\right)/\gamma_i^*\right| < \varepsilon_2$, where ε_2 is a prescribed bound, then $K_{opt} = K_i$, $p = p_i$. STOP.

4. Set $i = i + 1$ and $K_i = K_{i-1}$, then go to Step 2.

Remark 3.3 From [2], given $\mathcal{S}_\mathcal{I}$ positive and suppose $K_1 \leq\leq 0$, $\mathcal{S}_\mathcal{C}$ is positive and robustly stable for any $A \in [\underline{A}, \overline{A}]$, $B \in [\underline{B}, \overline{B}]$, $C_z \in [\underline{C}_z, \overline{C}_z]$, $D_z \in [\underline{D}_z, \overline{D}_z]$ and $C \in [\underline{C}, \overline{C}]$ if and only if there exist a diagonal matrix $P > 0$, a scalar $\epsilon > 0$, and matrices $L \leq\leq 0$ and $X \leq\leq 0$ such that

$$\Psi(X) \triangleq \begin{bmatrix} \bar{A}^T P + P\bar{A} + \epsilon X^T X - X^T L\underline{C} - \underline{C}^T L^T X & P\underline{B} + \underline{C}^T L^T \\ * & -\epsilon I \end{bmatrix} < 0, \quad (3.35)$$

$$\left[\epsilon\underline{A} + \overline{B}L\overline{C}\right]_{ij} \geq\geq 0,\ 1 \leq i \neq j \leq n, \quad (3.36)$$

$$\epsilon\underline{C}_z + \overline{D}_z L\overline{C} \geq\geq 0. \quad (3.37)$$

Under the condition, an output-feedback stabilizing control law can be obtained as

$$K_1 = \epsilon^{-1} L. \quad (3.38)$$

An iteration algorithm can be constructed to obtain the controller matrix K_1 by solving the conditions in (3.35)–(3.37), but omitted here for brevity [2].

Remark 3.4 The parameter γ is optimized iteratively. Notice that $\gamma_{i+1}^* \leq \gamma_i^*$ since the corresponding parameters obtained in Step 4 will be utilized as the initial conditions to derive a smaller γ. Therefore, the convergence of the iterative process is naturally guaranteed.

3.3 Special Case: $C = I$

When $C = I$, the static output-feedback controller synthesis problem is reduced to a state-feedback controller synthesis problem. Different from the approach developed earlier in this chapter, here an analytical method is established for the synthesis of the state-feedback controller for SIMO interval positive systems. The result is then extended to the MIMO case. In addition, an L_1-induced sparse controller is designed for positive systems.

3.3.1 *Controller Synthesis for SIMO Systems*

In the following, we present a theorem which serves as a characterization of the existence of a state-feedback controller.

Theorem 3.5 *Given $\mathcal{S}_\mathcal{I}$ with $C = I$ positive. $\mathcal{S}_\mathcal{C}$ with state-feedback is positive, robustly stable and satisfies $\|z\|_{L_1} < \gamma \|w\|_{L_1}$ under $x(0) = 0$ if and only if there exists a matrix K satisfying that*

$$A + BK \text{ is Metzler}, \tag{3.39}$$
$$C_z + D_z K \geq\geq 0, \tag{3.40}$$
$$\alpha(A + BK) < 0, \tag{3.41}$$
$$\|D_{zw} - (C_z + D_z K)(A + BK)^{-1} B_w\|_1 < \gamma. \tag{3.42}$$

Proof From (3.39)–(3.41), it is easy to see that $\mathcal{S}_\mathcal{C}$ with state-feedback is positive and stable.

The L_1-induced norm of system (3.20) with $C = I$ is

$$\|\tilde{\mathfrak{I}}\|_{L_1 \to L_1} = \|D_{zw} - (C_z + D_z K)(A + BK)^{-1} B_w\|_1. \tag{3.43}$$

From (3.42), we have $\|z\|_{L_1} < \gamma \|w\|_{L_1}$ holds.

The necessity can be obtained by reversing the above procedure and the proof is omitted here. □

If the elements of $A + BK$ are sufficiently small, (3.41) is more likely to be satisfied and from (3.43) and Lemma 1.5, the L_1-induced norm of system (3.20) with $C = I$ will also be small when the elements of $A+BK$ and $C_z+D_z K$ are sufficiently small.

In fact, for SIMO systems, the problem of jointly minimizing the off-diagonal elements of $A + BK$ and all the elements of $C_z + D_z K$ can be analytically solved. To show this, we rewrite B and D_z as vectors given by

$$B = b = \begin{bmatrix} b_1 \\ b_2 \\ \vdots \\ b_n \end{bmatrix} \in [\underline{b}, \overline{b}], \quad D_z = d_z = \begin{bmatrix} d_{z1} \\ d_{z2} \\ \vdots \\ d_{zq} \end{bmatrix} \in [\underline{d}_z, \overline{d}_z],$$

where $b \in \bar{\mathbb{R}}^n_+$ and $d_z \in \bar{\mathbb{R}}^q_+$. In this case, K is a row vector and write

$$K = \begin{bmatrix} k_1 \; k_2 \; \ldots \; k_n \end{bmatrix}.$$

For the SIMO case, we provide the following theorem.

Theorem 3.6 *Given $\mathcal{S}_\mathcal{I}$ with $C = I$ positive and SIMO with $\bar{b} \gg 0$ or $\bar{d}_z \gg 0$. The optimal K such that the off-diagonal elements of Metzler matrix $A + bK$ and the elements of $C_z + d_z K \geq\geq 0$ for any $A \in [\underline{A}, \overline{A}]$, $b \in [\underline{b}, \overline{b}]$, $C_z \in [\underline{C}_z, \overline{C}_z]$ and $d_z \in [\underline{d}_z, \overline{d}_z]$ are jointly minimized is given by*

$$K^* = \begin{bmatrix} k_1^* \; k_2^* \; \ldots \; k_n^* \end{bmatrix} \tag{3.44}$$

where

$$k_j^* = -\min_{\substack{i=1,2,\ldots,n\\ h=1,2,\ldots,q}} \left\{ \frac{\underline{a}_{ij}}{\overline{b}_i} \; (i \neq j), \frac{\underline{c}_{zhj}}{\overline{d}_{zh}} \right\}. \tag{3.45}$$

Moreover,

(i) *if* $\alpha(\overline{A} + \underline{b}K^*) < 0$, *then* $K^* = \arg\min \|\overline{D}_{zw} - (\overline{C}_z + \underline{d}_z K)(\overline{A} + \underline{b}K)^{-1}\overline{B}_w\|_1$ *subject to* $\underline{A} + \overline{b}K$ *is Metzler and* $\underline{C}_z + \overline{d}_z K \geq\geq 0$.

(ii) *For any* $K \leq\leq 0$ *satisfying (3.39) and (3.40), we have*

$$\alpha(\underline{A} + \overline{b}K^*) \leq \alpha(A + bK^*) \leq \alpha(A + bK) \leq \alpha(\overline{A} + \underline{b}K). \tag{3.46}$$

Proof Observe that the constraints $A + bK$ is Metzler and $C_z + d_z K \geq\geq 0$ on k_j come from the jth column of the two matrices only. With $A = [a_{ij}] \in [\underline{A}, \overline{A}]$ is Metzler, $b = [b_i] \in [\underline{b}, \overline{b}] \geq\geq 0$, $C_z = [c_{zhj}] \in [\underline{C}_z, \overline{C}_z] \geq\geq 0$ and $d_z = [d_{zh}] \in [\underline{d}_z, \overline{d}_z] \geq\geq 0$, these constraints on k_j can be written explicitly as

$$\begin{aligned} a_{ij} + b_i k_j &\geq 0, \quad && i = 1, 2, \ldots, n, \; (i \neq j), \\ c_{zhj} + d_{zh} k_j &\geq 0, \quad && h = 1, 2, \ldots, q. \end{aligned} \tag{3.47}$$

To minimize over the uncertainty domain, the values k_j, $j = 1, 2, \ldots, n$ should be the smallest possible while $A + bK$ is Metzler and $C_z + d_z K$ remains non-negative. Therefore, (3.47) equals to

$$\begin{aligned} \underline{a}_{ij} + \overline{b}_i k_j &\geq 0, \quad && i = 1, 2, \ldots, n, \; (i \neq j), \\ \underline{c}_{zhj} + \overline{d}_{zh} k_j &\geq 0, \quad && h = 1, 2, \ldots, q. \end{aligned} \tag{3.48}$$

In fact, the constraints become

$$k_j \geq -\min_{\substack{i=1,2,\ldots,n\\ h=1,2,\ldots,q}} \left\{ \frac{\underline{a}_{ij}}{\overline{b}_i} \; (i \neq j), \frac{\underline{c}_{zhj}}{\overline{d}_{zh}} \right\}. \tag{3.49}$$

Thus, to minimize the off-diagonal elements of $A + bK$ and all the elements of $C_z + d_z K$, the choice of k_j should be the minimum value subject to constraints in (3.49) which is given by k_j^* in (3.45). Hence, the first part of the result has been established.

When $\alpha(\overline{A} + \underline{b}K^*) < 0$, the system is robustly stable and $-(\overline{A} + \underline{b}K^*)^{-1} \geq\geq 0$. For any $K \geq\geq K^*$ satisfying $\alpha(\overline{A} + \underline{b}K) < 0$, it follows from Lemma 1.6 that $0 \leq\leq -(\overline{A} + \underline{b}K^*)^{-1} \leq\leq -(\overline{A} + \underline{b}K)^{-1}$ holds. Since $0 \leq\leq \overline{C}_z + \underline{d}_z K^* \leq\leq \overline{C}_z + \underline{d}_z K$, we have, from Lemma 1.5, that

$$0 \leq\leq -(\overline{C}_z + \underline{d}_z K^*)(\overline{A} + \underline{b}K^*)^{-1} \leq\leq -(\overline{C}_z + \underline{d}_z K)(\overline{A} + \underline{b}K)^{-1}$$

and consequently,

$$0 \leq\leq \overline{D}_{zw} - (\overline{C}_z + \underline{d}_z K^*)(\overline{A} + \underline{b}K^*)^{-1} B_w \leq\leq \overline{D}_{zw} - (\overline{C}_z + \underline{d}_z K)(\overline{A} + \underline{b}K)^{-1} B_w.$$

The 1-norm of these matrices thus gives

$$\|\overline{D}_{zw} - (\overline{C}_z + \underline{d}_z K^*)(\overline{A} + \underline{b}K^*)^{-1}\overline{B}_w\|_1 \leq \|\overline{D}_{zw} - (\overline{C}_z + \underline{d}_z K)(\overline{A} + \underline{b}K)^{-1}\overline{B}_w\|_1.$$

As K^* gives the smallest K elementwise satisfying the constraints that $\underline{A} + \overline{b}K$ is Metzler and $\underline{C}_z + \overline{d}_z K \geq\geq 0$, the value of the L_1-induced performance corresponds to its minimal value.

Moreover, to minimize over the uncertainty domain, K should be the smallest possible while $A + bK$ is Metzler and $C_z + d_z K$ remains non-negative. Therefore, $K^* \leq\leq K \leq\leq 0$ holds when $A + bK$ and $C_z + d_z K$ approach the minimal values. Therefore, from Lemma 1.5, we have $\alpha(\underline{A} + \overline{b}K^*) \leq \alpha(A + bK^*) \leq \alpha(A + bK) \leq \alpha(\overline{A} + \underline{b}K)$. The proof is completed. □

If K satisfies (3.39) and (3.40) and $\alpha(A + bK) < 0$, from (3.46), we have $\alpha(A + bK^*) \leq \alpha(A + bK) < 0$. The convergence rate for continuous systems is characterized by the spectral abscissa and therefore K^* in (3.44) gives the fastest convergence rate. On the other hand, due to the uncertainty, the convergence rate of the closed-loop system with K^* is different when $A,\ b,\ C_z$ and d_z take different values. From (3.46), we have $\alpha(\underline{A} + \overline{b}K^*) \leq \alpha(A + bK^*)$ and therefore, the convergence rate of system $(\underline{A},\ \overline{b},\ \underline{C}_z,\ \overline{d}_z)$ under closed-loop control with controller K^* in (3.44) is the fastest over the whole uncertainty domain.

Remark 3.5 For $\mathcal{S_I}$ with $C = I$ positive and SIMO with $\bar{b} \gg 0$ or $\bar{d}_z \gg 0$, Theorem 3.6 gives a complete solution to the minimal L_1-induced performance problem when $\alpha(\overline{A} + \underline{b}K^*) < 0$. Notice that the optimal K in (3.44) gives the fastest convergence rate when the closed-loop system is robustly stable, but it does not minimize $\alpha(\overline{A} + \underline{b}K)$ due to the presence of the constraint $\underline{C}_z + \overline{d}_z K \geq\geq 0$.

Remark 3.6 For $\mathcal{S_I}$ with $C = I$ positive and SIMO with $\overline{d}_z \neq 0$ or $\overline{b}$ having more than one nonzero element, k_j can be computed optimally according to (3.45) with the corresponding term inside $\{\cdot\}$ disappears if $\overline{b}_i = 0$ or $\overline{d}_{zh} = 0$.

It can be easily seen that if $\overline{d}_z = 0$, then the constraint $\underline{C}_z + \overline{d}_z K \geq\geq 0$ is naturally satisfied. Here, we consider the following two cases when $\overline{d}_z = 0$.

Case 1 For $\mathcal{S_I}$ with $C = I$ positive and SIMO with $\overline{d}_z = 0$ and $\overline{b}$ having only one nonzero element, say, $\overline{b}_s \neq 0$, then k_s can be chosen such that $\underline{a}_{ss} + \overline{b}_s k_s$ is arbitrarily negative and the remaining k_j $(j \neq s)$ can be computed optimally according to

$$k_j^* = -\frac{\underline{a}_{sj}}{\overline{b}_s}.$$

Case 2 For $\mathcal{S}_{\mathcal{I}}$ with $C = I$ positive and SIMO with $\overline{d}_z = 0$ and $\overline{b}$ having more than one nonzero element, k_j can be computed optimally according to

$$k_j^* = -\min_{i=1,2,\ldots,n} \left\{ \frac{\underline{a}_{ij}}{\overline{b}_i} \ (i \neq j) \right\},$$

with the corresponding term inside $\{\cdot\}$ disappears if $\overline{b}_i = 0$.

Then, we have the following corollary based on the above discussion.

Corollary 3.1 *For $\mathcal{S}_{\mathcal{I}}$ with $C = I$ positive and SIMO with $\overline{d}_z = 0$, the optimal K obtained in Case 1 and Case 2 minimizes $\alpha(\overline{A} + \underline{b}K)$ subject to $\underline{A} + \overline{b}K$ being Metzler. If $\alpha(\overline{A} + \underline{b}K^*) < 0$, then the closed-loop system achieves the minimal L_1-induced performance.*

3.3.2 Controller Synthesis for MIMO Systems

Although Theorem 3.5 is valid for general MIMO systems, unfortunately, the problem of jointly minimizing the off-diagonal elements of Metzler matrix $A + BK$ and all the elements of $C_z + D_z K \geq\geq 0$ is not generally well-posed. In the following, for MIMO systems, we present a method to derive the minimal K under a structural constraint.

Suppose $K \in \mathbb{R}^{m\times n}$ takes the following structure:

$$K = L\Delta \tag{3.50}$$

where $L \triangleq \begin{bmatrix} l_1 \ l_2 \ \ldots \ l_n \end{bmatrix} \in \mathbb{R}_+^{m\times n}$ is a fixed 'direction matrix' with $l_j \geq\geq 0$, $j = 1, 2, \ldots, n$, being unit vectors, and $\Delta \triangleq \mathrm{diag}(\delta_1, \delta_2, \ldots, \delta_n)$ with δ_j to be determined.

For ease of reference, we write

$$B = \begin{bmatrix} \hat{b}_1^T \\ \hat{b}_2^T \\ \vdots \\ \hat{b}_n^T \end{bmatrix} \in [\underline{B}, \overline{B}], \quad D_z = \begin{bmatrix} \hat{d}_{z1}^T \\ \hat{d}_{z2}^T \\ \vdots \\ \hat{d}_{zq}^T \end{bmatrix} \in [\underline{D}_z, \overline{D}_z]$$

and there hold $\hat{b}_i^T l_j \geq 0$ and $\hat{d}_{zh}^T l_j \geq 0$.

The following proposition provides a characterization of the optimal structured K which jointly minimizing the off-diagonal elements of Metzler matrix $A + BK$ and all the elements of $C_z + D_z K \geq\geq 0$. The proof is similar to that of Theorem 3.6 and thus omitted here for brevity.

Proposition 3.1 *Given $\mathcal{S}_{\mathcal{I}}$ positive with $\overline{B}^T l_j \gg 0$ or $\overline{D}_z^T l_j \gg 0$. The optimal structured K in (3.50) such that the off-diagonal elements of Metzler matrix $A + BK$ and the elements of $C_z + D_z K \geq\geq 0$ for any $A \in [\underline{A}, \overline{A}]$, $B \in [\underline{B}, \overline{B}]$, $C_z \in [\underline{C}_z, \overline{C}_z]$ and $D_z \in [\underline{D}_z, \overline{D}_z]$ are jointly minimized has Δ given by*

$$\Delta^* = \text{diag}(\delta_1^*, \delta_2^*, \ldots, \delta_n^*)$$

where

$$\delta_j^* = - \min_{\substack{i=1,2,\ldots,n \\ h=1,2,\ldots,q}} \left\{ \frac{\underline{a}_{ij}}{\overline{\hat{b}}_i^T l_j} \ (i \neq j), \frac{\underline{c}_{zhj}}{\overline{\hat{d}}_{zh}^T l_j} \right\},$$

with the corresponding term inside $\{\cdot\}$ disappears if $\overline{\hat{b}}_i^T l_j = 0$ or $\overline{\hat{d}}_{zh}^T l_j = 0$.

Remark 3.7 It is noted that the L_1-induced performance of an MIMO system is minimized under the structured K when $\alpha(\overline{A} + \underline{B}L\Delta^*) < 0$. However, if the closed-loop system is unstable, one has to choose another direction matrix L to search a possible K to stabilize the system. The selection of an appropriate direction matrix L to minimize the L_1-induced performance is much challenging and represents a future research direction.

3.3.3 *Sparse Controller Synthesis*

This subsection deals with the robust sparse state-feedback stabilization problem for positive systems with interval uncertainties. Finding sparse vectors is important in many applications such as in parameter estimation or identification, signal processing or model reduction. A vector or signal is said to be sparse, if most of its entries are zero. The ℓ_0-norm quantifies sparsity by counting the number of non-zero entries in a vector or signal. In detail, for a vector $x = [x_1, \ldots, x_n]^T \in \mathbb{R}^n$, it is called sparse if its ℓ_0-norm is small compared to the dimension of the vector, where ℓ_0-norm of x is defined as

$$\|x\|_0 = \sum_{i=1}^{n} |\text{sign}(x_i)|.$$

The problem of L_1-induced sparse controller design (L1SCD) is formulated as follows.

Problem L1SCD: *Given a positive system*

$$\mathcal{S} : \begin{cases} \dot{x}(t) = Ax(t) + Bu(t) + B_w w(t), \\ y(t) = Cx(t) + Du(t) + D_w w(t), \end{cases} \tag{3.51}$$

where $A \in [\underline{A}, \overline{A}]$, $B \in [\underline{B}, \overline{B}]$, $B_w \in [\underline{B}_w, \overline{B}_w]$, $C \in [\underline{C}, \overline{C}]$, $D \in [\underline{D}, \overline{D}]$ *and* $D_w \in [\underline{D}_w, \overline{D}_w]$*, find a state-feedback controller* $u(t) = Kx(t)$ *such that*

1. *the closed-loop system*

$$\begin{cases} \dot{x}(t) = (A + BK)x(t) + B_w w(t), \\ y(t) = (C + DK)x(t) + D_w w(t), \end{cases} \tag{3.52}$$

   is positive and robustly stable.
2. *the* ℓ_0*-norm of the controller gain* K *is minimized subject to the* L_1*-induced performance, that is,*

$$\begin{aligned} &\min \|K\|_0 \\ &\text{subject to } \|y\|_{L_1} < \gamma \|w\|_{L_1}. \end{aligned} \tag{3.53}$$

This is a common sense approach which simply seeks the sparsest controller gain K satisfying the constraint. However, the optimization problem is non-convex and NP-hard. From Lemma 1.7, we know that the ℓ_0-optimization problem (3.53) can be relaxed by the following ℓ_1-minimization problem:

$$\begin{aligned} &\min \|\text{vec}(K)\|_1 \\ &\text{subject to } \|y\|_{L_1} < \gamma \|w\|_{L_1}. \end{aligned} \tag{3.54}$$

Before solving the ℓ_1-minimization problem (3.54), a necessary and sufficient condition is first presented for the existence of a solution to the L_1-induced state-feedback stabilization problem. The proof is similar to that of Theorem 3.4 and is hence omitted.

Theorem 3.7 *Suppose* $K \leq\leq 0$. *The closed-loop system (3.52) is positive, robustly stable and satisfies* $\|z\|_{L_1} < \gamma \|w\|_{L_1}$ *for any* $A \in [\underline{A}, \overline{A}]$, $B \in [\underline{B}, \overline{B}]$, $B_w \in [\underline{B}_w, \overline{B}_w]$, $C \in [\underline{C}, \overline{C}]$, $D \in [\underline{D}, \overline{D}]$ *and* $D_w \in [\underline{D}_w, \overline{D}_w]$ *if and only if there exist a matrix* K *and a vector* $p \geq\geq 0$ *satisfying*

$$\underline{A} + \overline{B}K \text{ is Metzler}, \tag{3.55}$$

$$\underline{C} + \overline{D}K \geq\geq 0, \tag{3.56}$$

$$\mathbf{1}^T(\overline{C} + \underline{D}K) + p^T(\overline{A} + \underline{B}K) \ll 0, \tag{3.57}$$

$$p^T\overline{B}_w + \mathbf{1}^T\overline{D}_w - \gamma\mathbf{1}^T \ll 0. \tag{3.58}$$

Note that the Lyapunov vector p is coupled with the controller matrix K in (3.57), which cannot be easily solved. However, when matrix K is fixed, (3.57) turns out to be linear with respect to the remaining variables. Therefore, a natural way is to fix K, and solve (3.55)–(3.58) by linear programming. Thus, the following iterative algorithm can be proposed to solve Problem L1SCD.

Algorithm

- Step 1. Set $i = 1$. Select an initial matrix $K_1 \leq\leq 0$ such that the positive system

$$\begin{cases} \dot{x}(t) = Ax(t) + Bu(t) + B_w w(t), \\ y(t) = Cx(t) + Du(t) + D_w w(t), \end{cases} \tag{3.59}$$

  with

$$u(t) = K_1 x(t) \tag{3.60}$$

  is positive and robustly stable.
- Step 2. For fixed $K_i \leq\leq 0$, solve the following feasibility problem for $p_i \geq\geq 0$.
  FP: Find p_i subject to the following constraints:

$$\begin{aligned} \mathbf{1}^T(\overline{C} + \underline{D}K_i) + p_i^T(\overline{A} + \underline{B}K_i) &\ll 0, \\ p_i^T\overline{B}_w + \mathbf{1}^T\overline{D}_w - \gamma\mathbf{1}^T &\ll 0. \\ p_i &\geq\geq 0. \end{aligned}$$

  Denote p_i as the solution to the feasibility problem.
- Step 3. For fixed p_i, solve the following optimization problem for K_i.
  OP: Minimize $\|\text{vec}(K)\|_1$ subject to the following constraints:

$$\begin{aligned} \underline{A} + \overline{B}K_i &\text{ is Metzler}, \\ \underline{C} + \overline{D}K_i &\geq\geq 0, \\ \mathbf{1}^T(\overline{C} + \underline{D}K_i) + p_i^T(\overline{A} + \underline{B}K_i) &\ll 0. \end{aligned}$$

  Denote K_i^* as the solution to the optimization problem.
- Step 4. If $\left|\left(\|\text{vec}(K_i^*)\|_1 - \|\text{vec}(K_{i-1}^*)\|_1\right) / \|\text{vec}(K_i^*)\|_1\right| < \varepsilon_2$, where ε_2 is a prescribed tolerance, STOP; else set $i = i + 1$ and $K_i = K_{i-1}$, then go to Step 2.

Remark 3.8 The initial matrix K_1 can be viewed as a state-feedback controller matrix, and can be constructed by existing convex optimization approaches. From [3], we know that system (3.59) with (3.60) is positive and robustly stable if and only if there exist matrices $P \triangleq \text{diag}(p_1, p_2, \ldots, p_n)$ and $Q \triangleq [q_{ij}] \in \mathbb{R}^{l \times n}$ such that

$$\begin{aligned} &P\overline{A}^T + Q^T\overline{B}^T + \overline{A}P + \overline{B}Q < 0, \\ &\underline{a}_{ij}p_j + \sum_{z=1}^{l} \overline{b}_{iz}q_{zj} \geq 0, \ (1 \leq i \neq j \leq n) \\ &\underline{c}_{ij}p_j + \sum_{z=1}^{l} \overline{d}_{iz}q_{zj} \geq 0. \end{aligned}$$

Under this condition, an initial choice of K can be given by $K_1 = QP^{-1}$.

3.4 Dynamic Output-Feedback Controller Design

In this section, the dynamic output-feedback stabilization problem is studied for interval positive systems. The general controller structure under consideration is of the form

$$\begin{cases} \dot{\hat{x}}(t) = F\hat{x}(t) + Gy(t) \\ u(t) = K\hat{x}(t) \end{cases} \tag{3.61}$$

where $F \in \mathbb{R}^{n\times n}$, $G \in \mathbb{R}^{n\times r}$, and $K \in \mathbb{R}^{p\times n}$ are the controller matrices to be designed.

Define $e = x(t) - \hat{x}(t)$ and $\xi(t) = [x^T(t) \quad e^T(t)]^T$, then it follows from (3.15) and (3.61) that the augmented system can be described by

$$\begin{cases} \dot{\xi}(t) = A_\xi \xi(t) + B_{\xi w} w(t), \\ z(t) = C_\xi \xi(t) + D_{\xi w} w(t), \end{cases} \tag{3.62}$$

where

$$A_\xi = \begin{bmatrix} A + BK & -BK \\ A - GC + BK - F & F - BK \end{bmatrix}, \quad B_\xi = \begin{bmatrix} B_w \\ B_w \end{bmatrix}, \tag{3.63}$$

$$C_\xi = \begin{bmatrix} C_z + D_z K & -D_z K \end{bmatrix}, \quad D_{\xi w} = D_{zw}. \tag{3.64}$$

In the following, the problem of L_1-induced dynamic output-feedback controller design (L1DOFCD) is formulated.

Problem L1DOFCD: *Given $\mathcal{S}_\mathcal{I}$ positive. Find a dynamic output-feedback controller* (3.61) *with Metzler F, $G \geq\geq 0$ and $K \leq\leq 0$ such that the closed-loop system* (3.62) *is positive, robustly stable, and satisfies the L_1-induced performance in* (3.3) *under* $x(0) = 0$.

In this section, we restrict $e(t) \geq 0$. The positivity specification on the error signals $e(t)$ facilitates the synthesis of the desired dynamic output-feedback controller. Although this specification may cause some conservatism, the non-negativeness of the error signals $e(t)$ will not affect that of the estimation $\hat{x}(t)$. We are now in a position to establish conditions to Problem L1DOFCD.

Theorem 3.8 *Given $\mathcal{S}_\mathcal{I}$ positive and suppose F is Metzler, $G \geq\geq 0$, $K \leq\leq 0$. The closed-loop system (3.62) is positive, robustly stable and satisfies $\|z\|_{L_1} < \gamma\|w\|_{L_1}$ for any $A \in [\underline{A}, \overline{A}]$, $B \in [\underline{B}, \overline{B}]$, $B_w \in [\underline{B}_w, \overline{B}_w]$, $C_z \in [\underline{C}_z, \overline{C}_z]$, $D_z \in [\underline{D}_z, \overline{D}_z]$, $D_{zw} \in [\underline{D}_{zw}, \overline{D}_{zw}]$ and $C \in [\underline{C}, \overline{C}]$ if and only if there exist vectors $p_1 \geq\geq 0$, $p_2 \geq\geq 0$ satisfying*

$$\underline{A} + \overline{B}K \text{ is Metzler}, \quad (3.65)$$
$$\underline{A} - G\overline{C} + \overline{B}K - F \geq\geq 0, \quad (3.66)$$
$$\underline{C}_z + \overline{D}_z K \geq\geq 0, \quad (3.67)$$
$$\mathbf{1}^T(\overline{C}_z + \underline{D}_z K) + p_1^T(\overline{A} + \underline{B}K) + p_2^T(\overline{A} - G\underline{C} + \underline{B}K - F) \ll 0, \quad (3.68)$$
$$-\mathbf{1}^T\overline{D}_z K - p_1^T\overline{B}K + p_2^T(F - \overline{B}K) \ll 0, \quad (3.69)$$
$$p_1^T\overline{B}_w + p_2^T\overline{B}_w + \mathbf{1}^T\overline{D}_{zw} - \gamma\mathbf{1}^T \ll 0. \quad (3.70)$$

Proof Combining (3.65)–(3.67) with $G \geq\geq 0$, $K \leq\leq 0$ yields the following: for any $A \in [\underline{A}, \overline{A}]$, $B \in [\underline{B}, \overline{B}]$, $C_z \in [\underline{C}_z, \overline{C}_z]$, $D_z \in [\underline{D}_z, \overline{D}_z]$ and $C \in [\underline{C}, \overline{C}]$,

$$A + BK \geq\geq \underline{A} + \overline{B}K \text{ is Metzler}, \quad (3.71)$$
$$A - GC + BK - F \geq\geq \underline{A} - G\overline{C} + \overline{B}K - F \geq\geq 0, \quad (3.72)$$
$$C_z + DK \geq\geq \underline{C}_z + \overline{D}_z K \geq\geq 0, \quad (3.73)$$

which, together with $K \leq\leq 0$, imply

$$A_\xi = \begin{bmatrix} A + BK & -BK \\ A - GC + BK - F & F - BK \end{bmatrix} \text{ is Metzler}, \quad B_\xi = \begin{bmatrix} B_w \\ B_w \end{bmatrix} \geq\geq 0,$$
$$C_\xi = \begin{bmatrix} C_z + D_z K & -D_z K \end{bmatrix} \geq\geq 0, \quad D_{\xi w} = D_{zw} \geq\geq 0.$$

Therefore, we have that the closed-loop system (3.62) being a positive system.

By Theorem 3.3, inequalities (3.68)–(3.70) imply that the closed-loop system (3.62) is robustly stable and satisfies $\|z\|_{L_1} < \gamma\|w\|_{L_1}$. This proves sufficiency.

The necessity can be obtained by reversing the above procedure and the proof is omitted here. □

Remark 3.9 Similar to the static output-feedback case, when B and D_z have no uncertainty, the constraint $K \leq\leq 0$ can be removed.

We know that many dynamic output-feedback synthesis problems can be reformulated as static output-feedback controller design problems. Therefore, an iterative algorithm similar to Algorithm L1SOFCD can be proposed to solve the conditions in Theorem 3.8, but omitted here for brevity.

It is worth noting that the initial choice of matrix $\tilde{K}$ is important and it can be viewed as a static output-feedback gain matrix. For a positive system in (3.15) with controller (3.61), assume $w = 0$ and we have the closed-loop system

$$\dot{\xi}(t) = \begin{bmatrix} A + BK & -BK \\ A - GC + BK - F & F - BK \end{bmatrix} \xi(t). \quad (3.74)$$

It is noted that system (3.74) is robustly stable for any $A \in [\underline{A}, \overline{A}]$, $B \in [\underline{B}, \overline{B}]$, and $C \in [\underline{C}, \overline{C}]$ if and only if the following system is stable:

$$\dot{\xi}(t) = \begin{bmatrix} \overline{A} + \underline{B}K & -\overline{B}K \\ \overline{A} - G\underline{C} + \underline{B}K - F & F - \overline{B}K \end{bmatrix} \xi(t), \tag{3.75}$$

which can be transformed to be

$$\dot{\xi}(t) = (\tilde{A} + \tilde{B}\tilde{K}\tilde{C})\xi(t), \tag{3.76}$$

where

$$\tilde{A} = \begin{bmatrix} \overline{A} & 0 \\ \overline{A} & 0 \end{bmatrix}, \ \tilde{B} = \begin{bmatrix} 0 & \underline{B} - \overline{B} & 0 & \underline{B} \\ 0 & \underline{B} - \overline{B} & I & \underline{B} \end{bmatrix},$$
$$\tilde{K} = \begin{bmatrix} F & G & 0 & 0 \\ -K & 0 & 0 & 0 \\ 0 & 0 & F & G \\ 0 & 0 & -K & 0 \end{bmatrix}, \ \tilde{C} = \begin{bmatrix} 0 & -I \\ 0 & 0 \\ -I & I \\ -\underline{C} & 0 \end{bmatrix}. \tag{3.77}$$

The initial matrix can be given by the following proposition.

Proposition 3.2 *Given $\mathcal{S}_{\mathcal{I}}$ positive and suppose F_1 is Metzler, $G_1 \geq\geq 0$, $K_1 \leq\leq 0$. System (3.15) is output-feedback stabilizable by controller (3.61) if and only if there exist a matrix $P > 0$, a scalar $\epsilon > 0$, and matrices L and X such that*

$$L = \begin{bmatrix} L_{11} & L_{12} & 0 & 0 \\ L_{21} & 0 & 0 & 0 \\ 0 & 0 & L_{11} & L_{12} \\ 0 & 0 & L_{21} & 0 \end{bmatrix},$$
$$\Psi(X) \triangleq \begin{bmatrix} \tilde{A}^T P + P\tilde{A} + \epsilon X^T X - X^T L\tilde{C} - \tilde{C}^T L^T X & P\tilde{B} + \tilde{C}^T L^T \\ * & -\epsilon I \end{bmatrix} < 0,$$
$$\left[\epsilon\underline{A} - \overline{B}L_{21}\right]_{ij} \geq 0, \ 1 \leq i \neq j \leq n,$$
$$\epsilon\underline{A} - L_{12}\overline{C} - \overline{B}L_{21} - L_{11} \geq\geq 0,$$
$$\epsilon\underline{C}_z - \overline{D}_z L_{21} \geq\geq 0,$$

where L_{11} is Metzler, *$L_{12} \geq\geq 0$, $L_{21} \geq\geq 0$ and $\tilde{A}$, $\tilde{B}$, $\tilde{C}$, $\tilde{K}$ are defined in (3.77). Under the condition, an output-feedback stabilizing control law can be obtained as*

$$\tilde{K} = \epsilon^{-1} L.$$

3.5 Numerical Examples

In this section, some examples are provided to illustrate the effectiveness of the proposed approach.

Example 3.1 To demonstrate the effectiveness of the static output-feedback controller design method in Theorem 3.4 and the analytical method proposed in Theorem 3.6 for the special case of $C = I$, the following positive system is considered:

$$\begin{cases} \dot{x}(t) = Ax(t) + Bu(t) + B_w w(t), \\ z(t) = C_z x(t) + D_z u(t), \\ y(t) = Cx(t), \end{cases} \tag{3.78}$$

where $A \in [\underline{A}, \overline{A}]$, $B \in [\underline{B}, \overline{B}]$, $B_w \in [\underline{B}_w, \overline{B}_w]$, $C_z \in [\underline{C}_z, \overline{C}_z]$, $D_z \in [\underline{D}_z, \overline{D}_z]$, C with

$$\underline{A} = \begin{bmatrix} -2.00 & 1.30 & 2.00 \\ 0.50 & -3.00 & 2.00 \\ 2.00 & 1.50 & -2.00 \end{bmatrix}, \ \overline{A} = \begin{bmatrix} -1.98 & 1.31 & 2.00 \\ 0.60 & -2.96 & 2.10 \\ 2.00 & 1.50 & -1.92 \end{bmatrix},$$

$$\underline{B} = \begin{bmatrix} 1.00 \\ 0.00 \\ 1.00 \end{bmatrix}, \ \overline{B} = \begin{bmatrix} 1.04 \\ 0.04 \\ 1.00 \end{bmatrix}, \ \underline{B}_w = \begin{bmatrix} 0.48 \\ 0.08 \\ 0.00 \end{bmatrix}, \ \overline{B}_w = \begin{bmatrix} 0.50 \\ 0.10 \\ 0.00 \end{bmatrix},$$

$$\underline{C}_z = \begin{bmatrix} 1.00 & 0.60 & 1.00 \\ 0.80 & 0.80 & 1.00 \end{bmatrix}, \ \overline{C}_z = \begin{bmatrix} 1.02 & 0.61 & 1.00 \\ 0.80 & 0.82 & 1.00 \end{bmatrix},$$

$$\underline{D}_z = \begin{bmatrix} 0.48 \\ 0.50 \end{bmatrix}, \ \overline{D}_z = \begin{bmatrix} 0.50 \\ 0.50 \end{bmatrix}, \ C = \begin{bmatrix} 0.20 & 0.00 & 0.00 \\ 0.00 & 0.50 & 0.20 \end{bmatrix}.$$

By solving the conditions in Theorem 3.4 using Algorithm L1SOFCD, we obtain an upper bound of γ^* as 0.1218 after 10 iterations and a feasible solution is achieved with

$$p = \begin{bmatrix} 0.2061 & 0.1880 & 0.9324 \end{bmatrix}^T,$$

which further yields the static output-feedback controller matrix as

$$K_{opt} = \begin{bmatrix} -7.9999 & -2.4000 \end{bmatrix}.$$

Figure 3.1 shows the variation of γ_i^* with iteration i. From Fig. 3.1 one can clearly see that γ_i^* is monotonically decreasing with respect to iteration i.

In the following, we investigate the case of $C = I$ and design a state feedback controller.

By resorting to the analytical method proposed in Sect. 3.3.1, we can easily obtain the controller gain matrix

$$K^* = \begin{bmatrix} -1.6000 & -1.2000 & -1.9230 \end{bmatrix}, \tag{3.79}$$

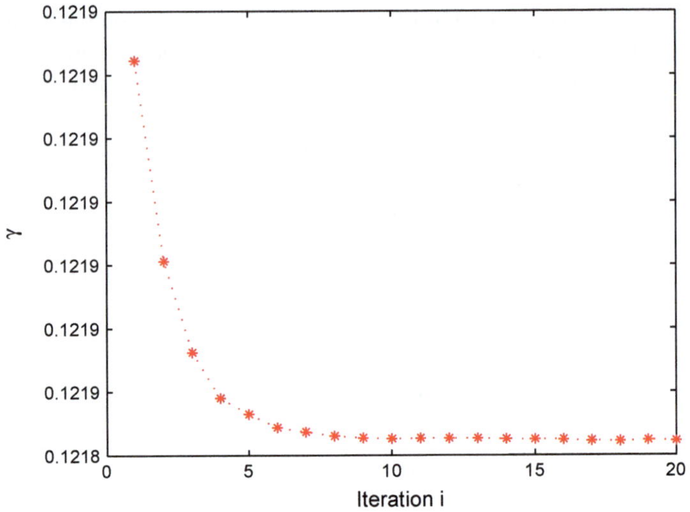

Fig. 3.1 Variation of γ_i^* with iteration i

which yields

$$A + BK = \begin{bmatrix} -2.49 \pm 0.01 & 0.25 \pm 0.05 & 1 \\ 0.55 \pm 0.05 & -2.98 \pm 0.02 & 2.05 \pm 0.05 \\ 1.5 & 1.05 \pm 0.05 & -2.96 \pm 0.04 \end{bmatrix},$$

$$C + DK = \begin{bmatrix} 0.76 \pm 0.01 & 0 & 0.5 \\ 0 & 0.51 \pm 0.01 & 0 \end{bmatrix}.$$

With the introduction of controller (3.79),

$$\alpha(\overline{A} + \underline{B}K^*) = -2.3965,$$
$$\|(\overline{C}_z + \underline{D}_z K^*)(\overline{A} + \underline{B}K^*)^{-1}\overline{B}_w\|_1 = 0.0580.$$

Moreover, we have

$$\alpha(\underline{A} + \overline{B}K^*) = -2.4750,$$
$$\|(\underline{C}_z + \overline{D}_z K^*)(\underline{A} + \overline{B}K^*)^{-1}\underline{B}_w\|_1 = 0.0390,$$

which correspond to the fastest convergence rate and the minimal L_1-induced performance of the closed-loop system, respectively.

To illustrate the disturbance attenuation performance, the external disturbance $w(t)$ is assumed to be

$$w(t) = \begin{cases} 300 + 200 \cos 2t, & t \leq 5, \\ 0, & \text{otherwise,} \end{cases}$$

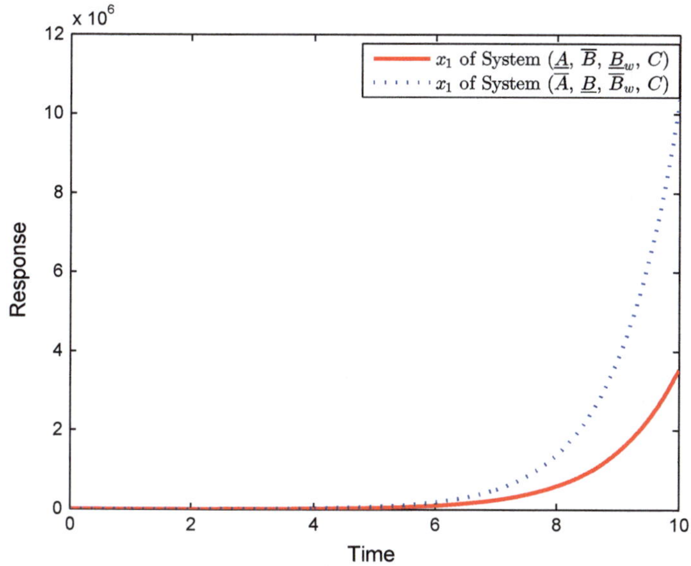

Fig. 3.2 Time response x_1 of open-loop system. Reproduced from Ref. [4] by permission of John Wiley and Sons Ltd

and the initial condition used in the simulation is

$$x(0) = \begin{bmatrix} 500\ 200\ 300 \end{bmatrix}^T.$$

The responses of the open-loop and the closed-loop system are shown in Figs. 3.2–3.10, from which we can see that the system can be stabilized by the designed controller.

Example 3.2 This example is given to illustrate the effectiveness of the proposed sparse controller design approach in Theorem 3.7. Consider the following positive system:

$$\begin{cases} \dot{x}(t) = Ax(t) + Bu(t) + B_w w(t), \\ y(t) = Cx(t) + Du(t) + D_w w(t), \end{cases} \tag{3.80}$$

where $A \in [\underline{A}, \overline{A}]$, $B \in [\underline{B}, \overline{B}]$, $B_w \in [\underline{B}_w, \overline{B}_w]$, $C \in [\underline{C}, \overline{C}]$, $D \in [\underline{D}, \overline{D}]$, $D_w \in [\underline{D}_w, \overline{D}_w]$ with

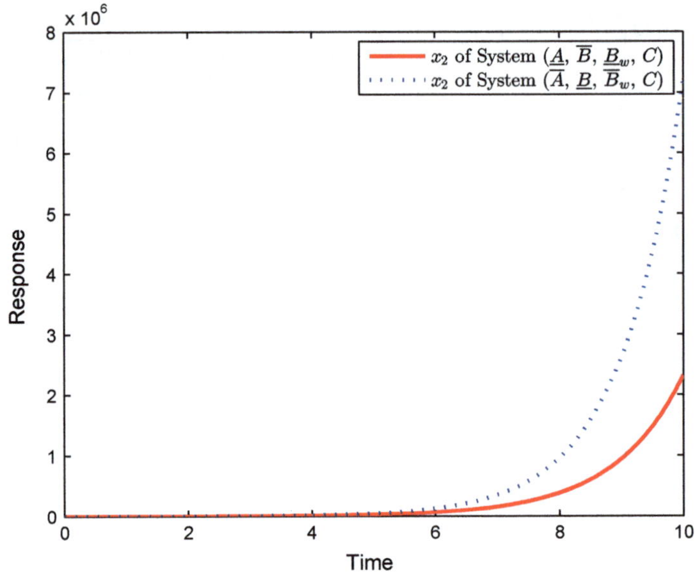

Fig. 3.3 Time response x_2 of open-loop system. Reproduced from Ref. [4] by permission of John Wiley and Sons Ltd

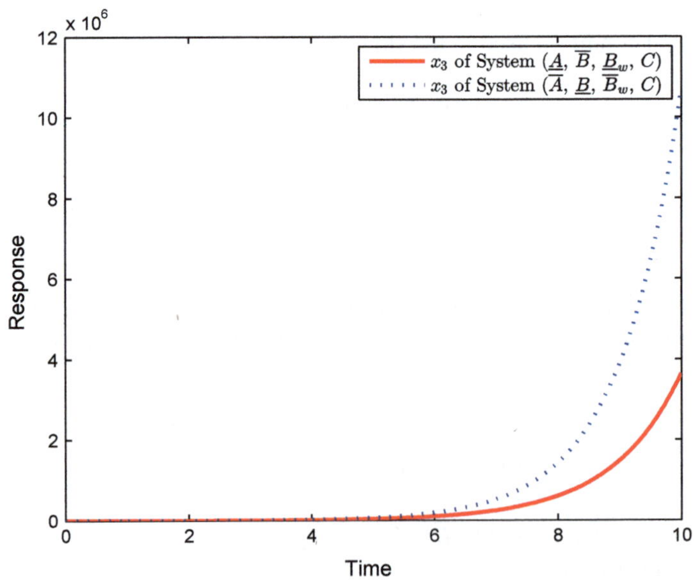

Fig. 3.4 Time response x_3 of open-loop system. Reproduced from Ref. [4] by permission of John Wiley and Sons Ltd

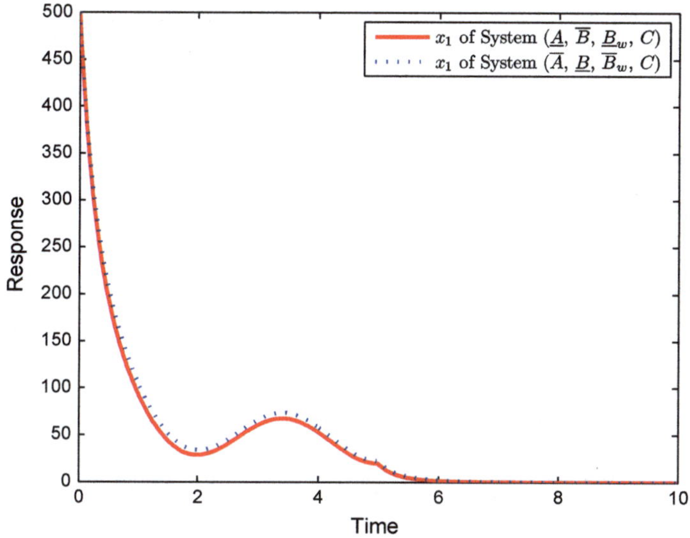

Fig. 3.5 Time response x_1 of closed-loop system. Reproduced from Ref. [4] by permission of John Wiley and Sons Ltd

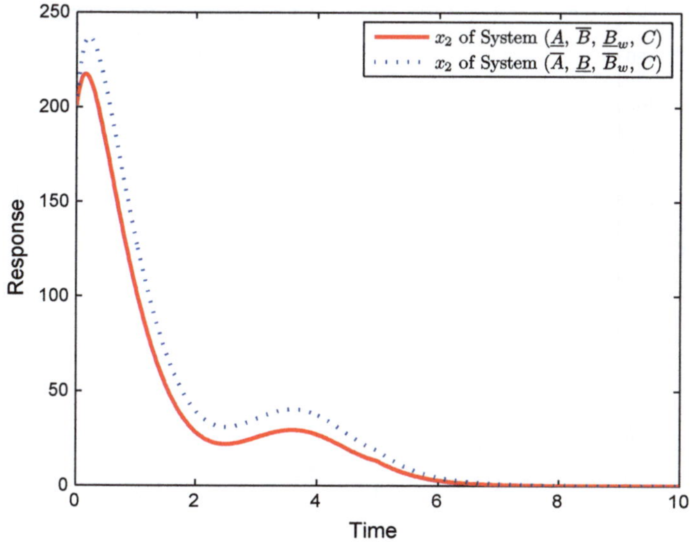

Fig. 3.6 Time response x_2 of closed-loop system. Reproduced from Ref. [4] by permission of John Wiley and Sons Ltd

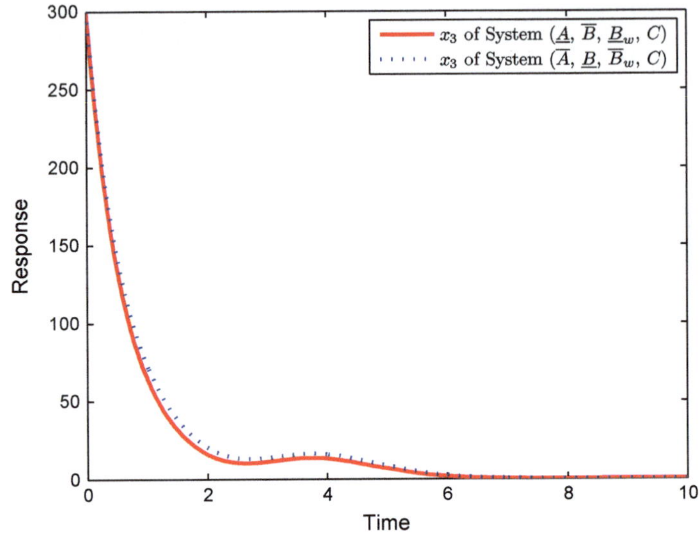

Fig. 3.7 Time response x_3 of closed-loop system. Reproduced from Ref. [4] by permission of John Wiley and Sons Ltd

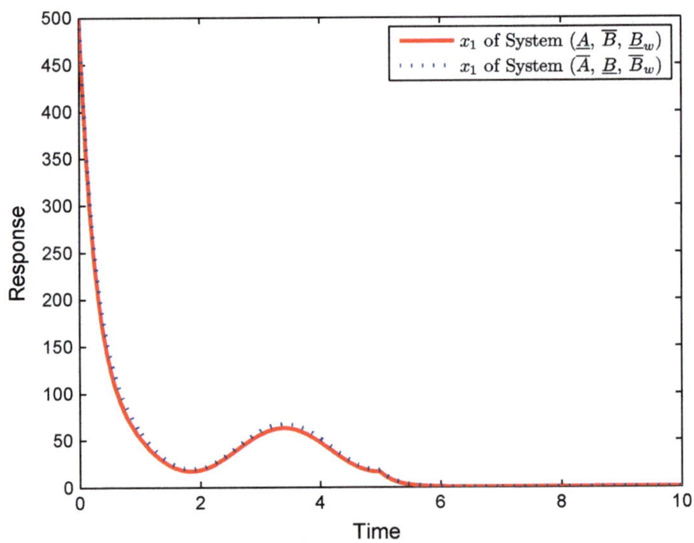

Fig. 3.8 Time response x_1 of closed-loop system with $C = I$. Reproduced from Ref. [4] by permission of John Wiley and Sons Ltd

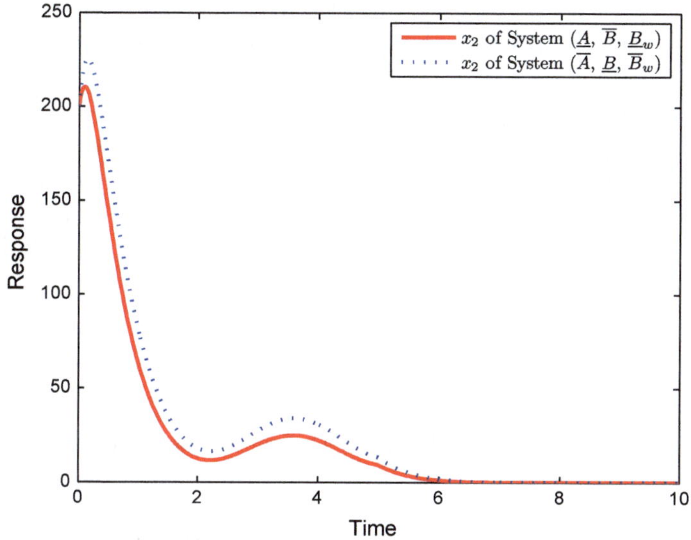

Fig. 3.9 Time response x_2 of closed-loop system with $C = I$. Reproduced from Ref. [4] by permission of John Wiley and Sons Ltd

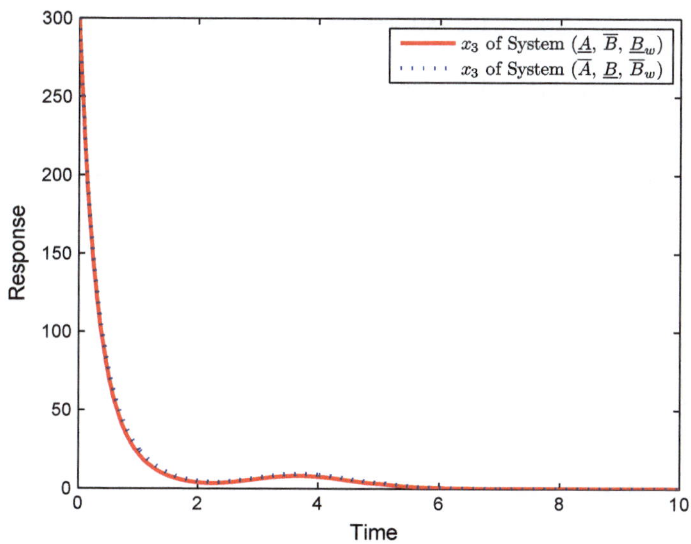

Fig. 3.10 Time response x_3 of closed-loop system with $C = I$. Reproduced from Ref. [4] by permission of John Wiley and Sons Ltd

$$\underline{A} = \begin{bmatrix} -1 & 0.5 & 0.2 & 0.5 \\ 0.5 & -2 & 0.5 & 0.1 \\ 0.1 & 0.2 & -1 & 0.3 \\ 0.5 & 0.5 & 1 & -0.5 \end{bmatrix}, \ \overline{A} = \begin{bmatrix} -0.98 & 0.51 & 0.2 & 0.5 \\ 0.5 & -1.97 & 0.52 & 0.1 \\ 0.1 & 0.2 & -1 & 0.3 \\ 0.52 & 0.5 & 1.01 & -0.5 \end{bmatrix},$$

$$\underline{B} = \begin{bmatrix} 5 \\ 4 \\ 0.8 \\ 4 \end{bmatrix}, \ \overline{B} = \begin{bmatrix} 5 \\ 4.01 \\ 0.82 \\ 4 \end{bmatrix}, \ \underline{B}_w = \begin{bmatrix} 0.01 \\ 0.01 \\ 0.08 \\ 0.01 \end{bmatrix}, \ \overline{B}_w = \begin{bmatrix} 0.02 \\ 0.012 \\ 0.1 \\ 0.01 \end{bmatrix},$$

$$\underline{C} = \begin{bmatrix} 0.6 & 0.0006 & 0.0006 & 0.0006 \end{bmatrix},$$

$$\overline{C} = \begin{bmatrix} 0.6 & 0.0008 & 0.0007 & 0.0006 \end{bmatrix},$$

$$\underline{D} = 0.5, \ \overline{D} = 0.52, \ \underline{D}_w = 0.01, \ \overline{D}_w = 0.015.$$

For $\gamma = 0.8$, by solving the conditions in Theorem 3.7 using Algorithm, we obtain the sparse controller gain

$$K^* = \begin{bmatrix} -0.1222 & 0.0000 & -0.0000 & 0.0000 \end{bmatrix}$$

after 10 iterations and a feasible solution is achieved with

$$p = \begin{bmatrix} 0.3980 & 1.0663 & 3.1810 & 2.5211 \end{bmatrix}.$$

Figure 3.11 shows the variation of $\|\text{vec}(K)\|_1$ with iteration i.

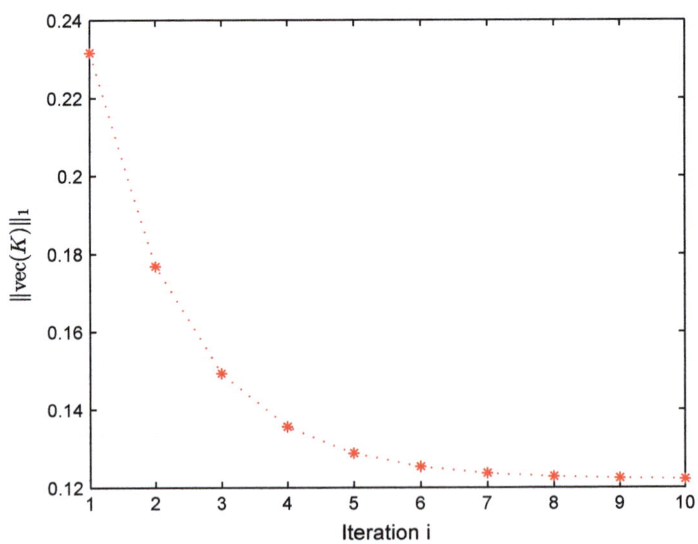

Fig. 3.11 Variation of $\|\text{vec}(K)\|_1$ with iteration i

To illustrate the disturbance attenuation performance, the external disturbance $w(t)$ is assumed to be

$$w(t) = \begin{cases} 100 + 20\cos t\,, & t \leq 10, \\ 0\,, & \text{otherwise}, \end{cases}$$

and the initial condition used in the simulation is

$$x(0) = \left[\, 500\ 200\ 300\ 400 \,\right]^T.$$

Figures 3.12–3.15 show the response of open-loop system and Figs. 3.16–3.19 show the state response of the closed-loop system, from which we can see that the system can be stabilized by the designed controller.

Example 3.3 In order to show the applicability of the dynamic output-feedback controller design method proposed in Theorem 3.8, the following continuous MIMO positive system is considered:

$$\begin{cases} \dot{x}(t) = Ax(t) + Bu(t) + B_w w(t), \\ z(t) = C_z x(t) + D_z u(t), \\ y(t) = Cx(t), \end{cases} \tag{3.81}$$

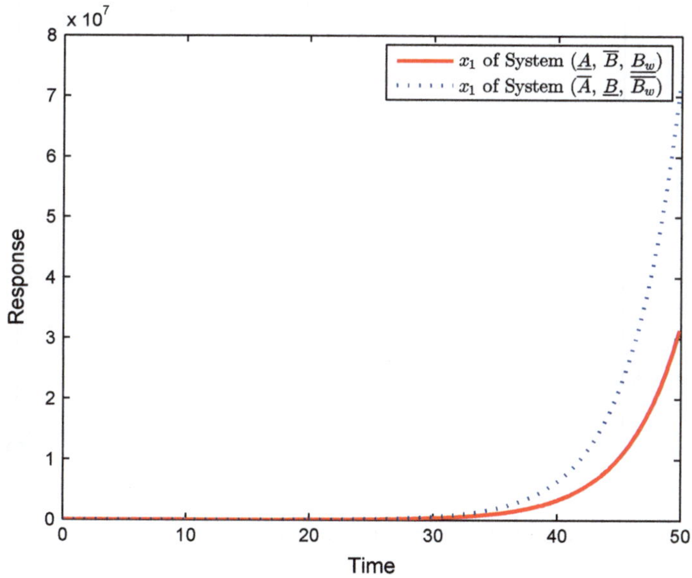

Fig. 3.12 Time response x_1 of open-loop system in Example 3.2

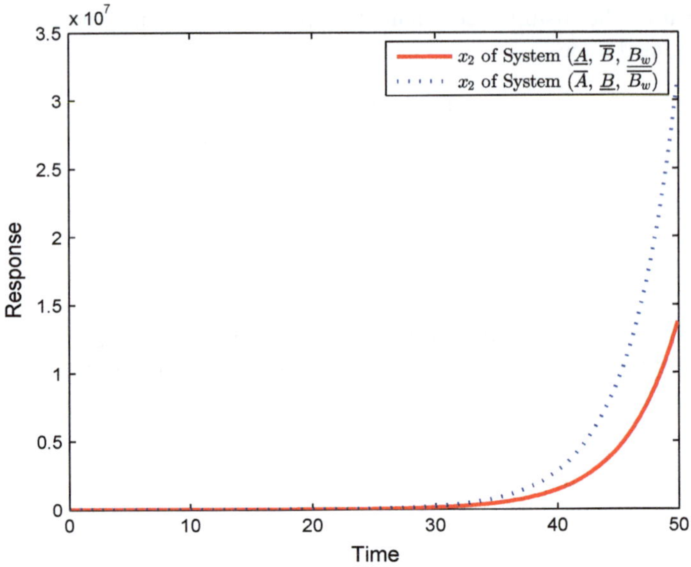

Fig. 3.13 Time response x_2 of open-loop system in Example 3.2

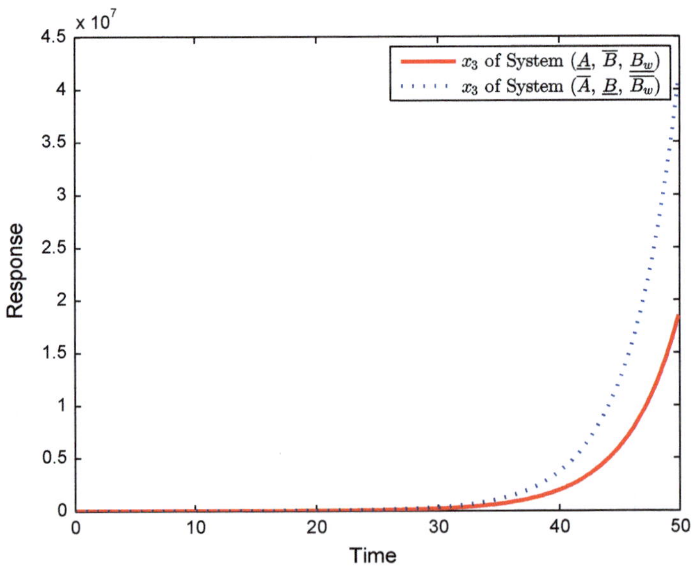

Fig. 3.14 Time response x_3 of open-loop system in Example 3.2

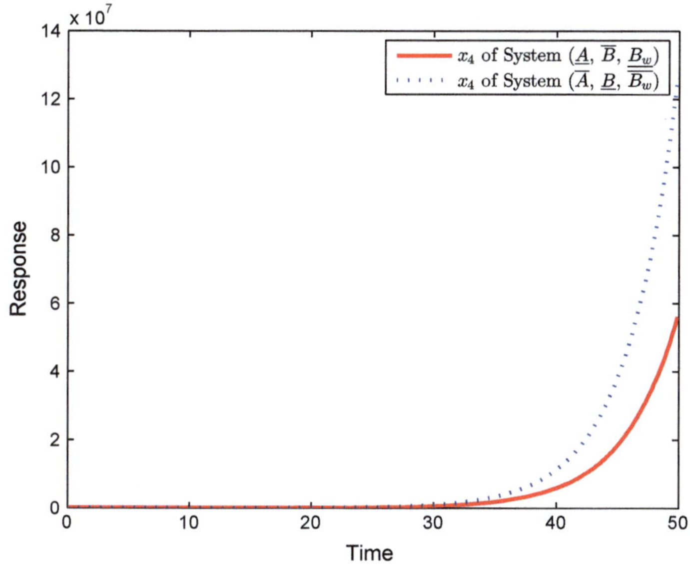

Fig. 3.15 Time response x_4 of open-loop system in Example 3.2

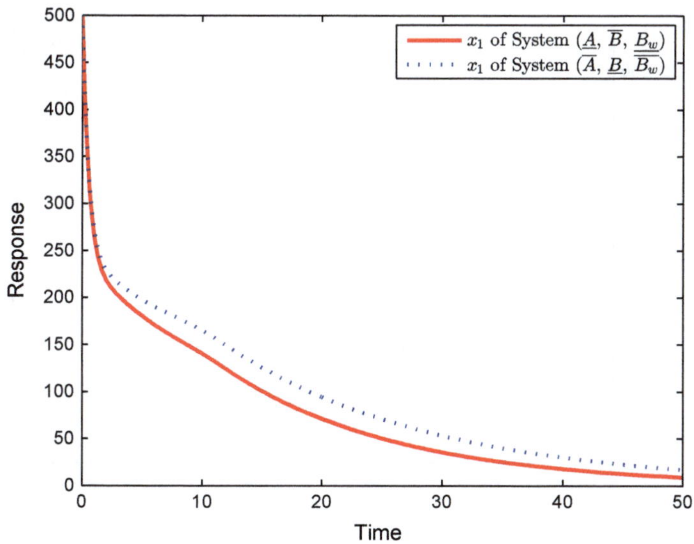

Fig. 3.16 Time response x_1 of closed-loop system in Example 3.2

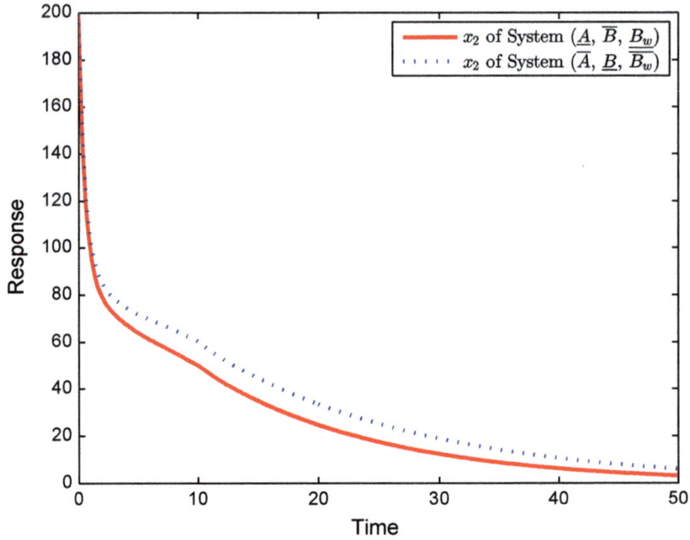

Fig. 3.17 Time response x_2 of closed-loop system in Example 3.2

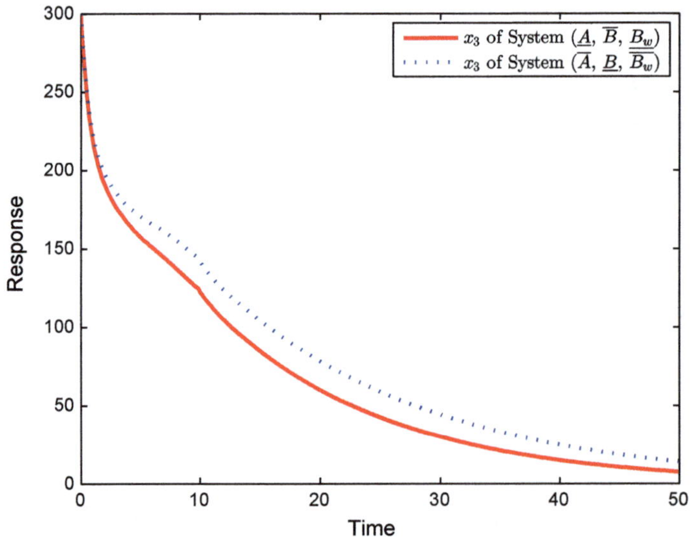

Fig. 3.18 Time response x_3 of closed-loop system in Example 3.2

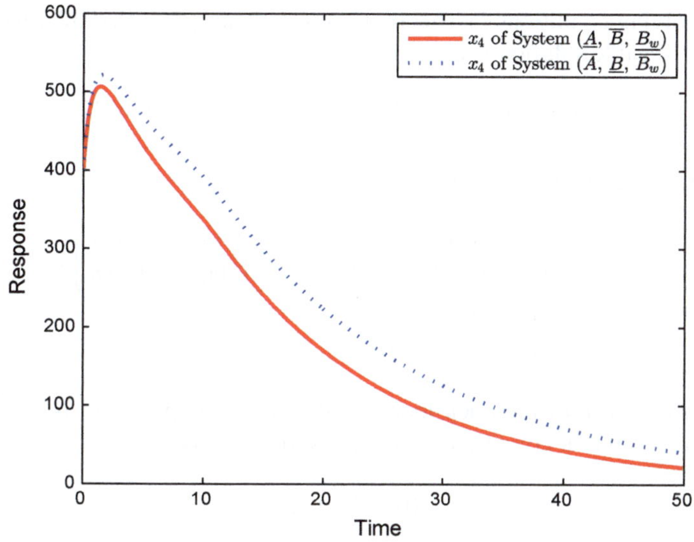

Fig. 3.19 Time response x_4 of closed-loop system in Example 3.2

where $A \in [\underline{A}, \overline{A}]$, $B \in [\underline{B}, \overline{B}]$, $B_w \in [\underline{B}_w, \overline{B}_w]$, $C_z \in [\underline{C}_z, \overline{C}_z]$, $D_z \in [\underline{D}_z, \overline{D}_z]$, $C \in [\underline{C}, \overline{C}]$ with

$$\underline{A} = \begin{bmatrix} -0.6000 & 0.5000 \\ 0.5000 & -0.4000 \end{bmatrix}, \overline{A} = \begin{bmatrix} -0.5900 & 0.5100 \\ 0.5100 & -0.3800 \end{bmatrix},$$
$$\underline{B} = \begin{bmatrix} 0.1000 & 0.0000 \\ 0.5000 & 0.0500 \end{bmatrix}, \overline{B} = \begin{bmatrix} 0.1100 & 0.0500 \\ 0.5000 & 0.0600 \end{bmatrix}, \underline{B}_w = \begin{bmatrix} 0.2000 \\ 0.1000 \end{bmatrix},$$
$$\overline{B}_w = \begin{bmatrix} 0.2100 \\ 0.1200 \end{bmatrix}, \underline{C}_z = \begin{bmatrix} 0.1100 & 0.1000 \end{bmatrix}, \overline{C}_z = \begin{bmatrix} 0.1200 & 0.1200 \end{bmatrix},$$
$$\underline{D}_z = \begin{bmatrix} 0.1000 & 0.1000 \end{bmatrix}, \overline{D}_z = \begin{bmatrix} 0.1200 & 0.1200 \end{bmatrix},$$
$$\underline{C} = \begin{bmatrix} 0.1000 & 0.1000 \\ 0.1000 & 0.5000 \end{bmatrix}, \overline{C} = \begin{bmatrix} 0.1100 & 0.1200 \\ 0.1000 & 0.5200 \end{bmatrix}.$$

In the following, we design a dynamic output-feedback controller for system (3.81). By implementing conditions in Proposition 3.2 via Yalmip, the initial matrices F_1, G_1, K_1 can be obtained as

$$F_1 = \begin{bmatrix} -0.9992 & 0.0001 \\ 0.0002 & -1.3703 \end{bmatrix}, G_1 = \begin{bmatrix} 2.5429 & 0.1934 \\ 1.2897 & 0.9671 \end{bmatrix},$$
$$K_1 = \begin{bmatrix} -0.4273 & -0.5596 \\ -0.4879 & -0.2700 \end{bmatrix}.$$

Then, by solving conditions in Theorem 3.8, we obtain $\gamma^* = 0.1032$ after 10 iterations and a feasible solution is achieved with

$$p_1 = \begin{bmatrix} 0.1245 & 0.1604 \end{bmatrix}^T, \ p_2 = \begin{bmatrix} 0.2056 & 0.1219 \end{bmatrix}^T,$$

which further yields the controller gain matrices as

$$F = \begin{bmatrix} -1.0585 & 0.0001 \\ 0.0000 & -1.8760 \end{bmatrix}, \ G = \begin{bmatrix} 3.4597 & 0.0004 \\ 0.0003 & 2.0988 \end{bmatrix},$$
$$K = \begin{bmatrix} -0.5342 & -0.7682 \\ -0.3824 & -0.0007 \end{bmatrix}.$$

The performance of the open-loop and the closed-loop system is evaluated via simulation. The initial condition used in the simulation is

$$x(0) = \begin{bmatrix} 20 & 15 \end{bmatrix}^T$$

and the external disturbance $w(t)$ is assumed to be

$$w(t) = \begin{cases} e^{-3t} + 10 \sin t, & t \leq 3, \\ 0, & \text{otherwise.} \end{cases}$$

Figures 3.20–3.23 show the response of the open-loop and the closed-loop system with the external disturbance $w(t)$ and it is noted that system can be stabilized by the designed controller.

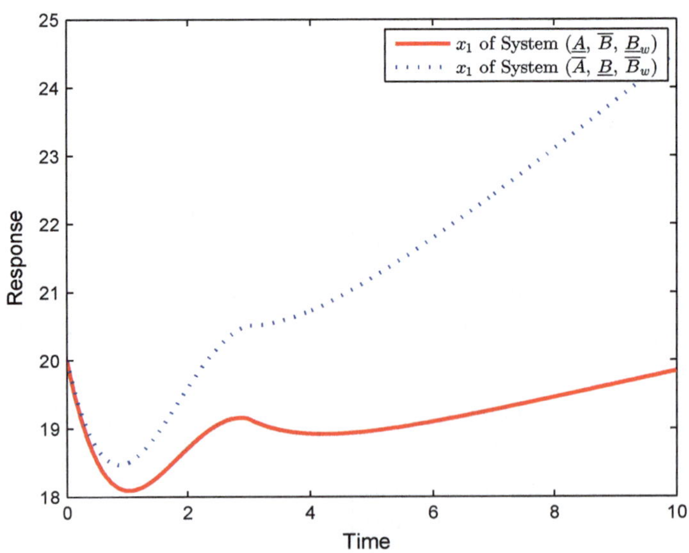

Fig. 3.20 Time response x_1 of open-loop system in Example 3.3

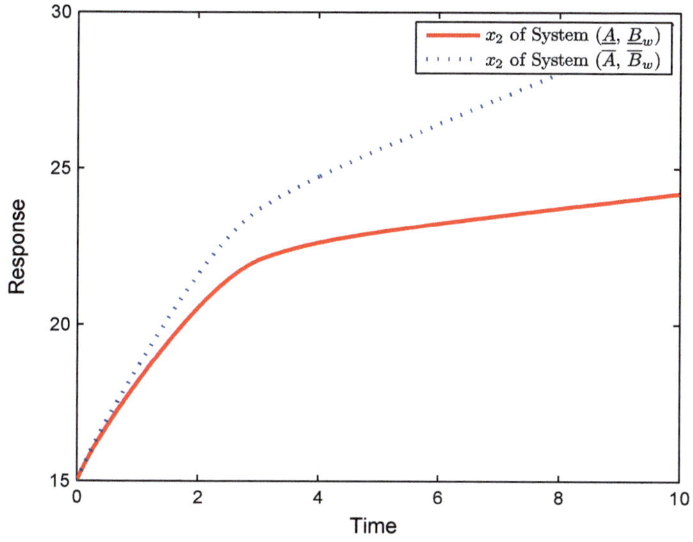

Fig. 3.21 Time response x_2 of open-loop system in Example 3.3

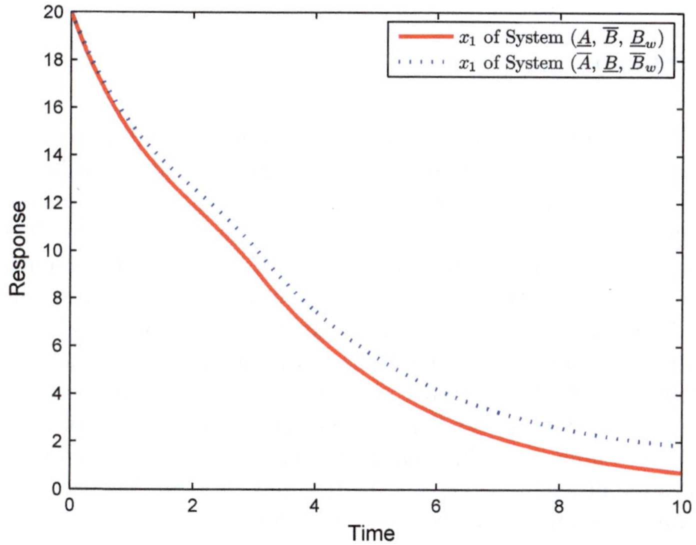

Fig. 3.22 Time response x_1 of closed-loop system in Example 3.3

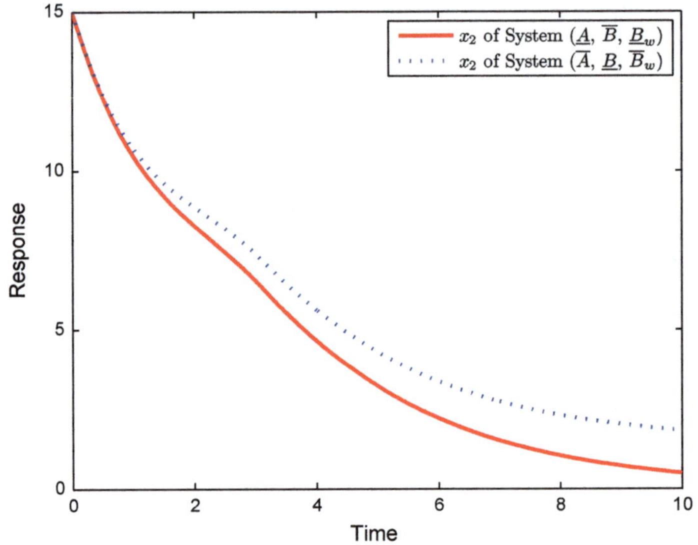

Fig. 3.23 Time response x_2 of closed-loop system in Example 3.3

3.6 Summary

In this chapter, we have studied the synthesis problem of L_1-induced output-feedback controllers for interval continuous positive systems. A method has been derived to compute the exact value of the L_1-induced norm for positive systems and a characterization has been proposed to ensure the asymptotic stability of the controlled system with a prescribed L_1-induced performance level. Based on the performance characterization, conditions for the existence of a desired static output-feedback controller have been established, and an iterative convex optimization algorithm has been developed for their solution. An analytical design approach has been put forward to investigate the controller synthesis for interval continuous positive systems with $C = I$. The sparse controller design problem has been formulated by minimizing the ℓ_0-norm of the controller gain K, which has been relaxed by an ℓ_1-minimization problem. Moreover, the resulting problem has been solved by an iterative convex optimization algorithm. In addition, the dynamic output-feedback controller has been designed for interval positive systems. Finally, three numerical examples have been provided to illustrate the effectiveness of the theoretical results.

References

1. Chellaboina VS, Haddad WM, Bernstein DS et al (2000) Induced convolution operator norms of linear dynamical systems. Math Control Signal Syst 13(3):216–239
2. Shu Z, Lam J (2008) An augmented system approach to static output-feedback stabilization with H_∞ performance for continuous-time plants. Int J Robust Nonlinear Control 19:768–785
3. Gao H, Lam J, Wang C et al (2005) Control for stability and positivity: equivalent conditions and computation. IEEE Trans Circuits Syst (II) 52(9):540–544
4. Chen X, Lam J, Li P et al (2014) Output-feedback control for continuous-time interval positive systems under L_1 performance. Asian J Control 16(6):1592–1601

Chapter 4
Positive State-Bounding Observer for Interval Positive Systems

In this chapter, the positive observer problem is investigated for interval positive systems under the L_1-induced performance. To estimate the state of positive systems, a pair of state-bounding positive observers is designed. A novel characterization is first proposed under which the augmented system is stable and satisfies the L_1-induced performance. Necessary and sufficient conditions are then presented to design the observers. The obtained results are expressed in terms of linear programming problems, and can be easily solved by standard software. In the end, a numerical example is presented to show the effectiveness of the derived design procedures.

This chapter is briefly organized as follows. The problem of state-bounding positivity preserving observers is formulated in Sect. 4.1. In Sect. 4.2, the positive state-bounding observer is designed for positive systems. To show the application of the theoretical results, an example is given in Sect. 4.3. Finally, the results are summarized in Sect. 4.4.

4.1 Problem Formulation

Consider a continuous-time positive linear system:

$$\begin{cases} \dot{x}(t) = Ax(t) + Bw(t), \\ y(t) = Cx(t) + Dw(t), \end{cases} \tag{4.1}$$

where $x(t) \in \mathbb{R}^n$, $w(t) \in \mathbb{R}^m$ and $y(t) \in \mathbb{R}^p$ are the system state, input and output, respectively; A, B, C and D are system matrices with compatible dimensions.

Here, the system matrices A, B, C and D belong to the following interval uncertainty domain:

$$A \in [\underline{A}, \overline{A}],\ \ B \in [\underline{B}, \overline{B}],\ \ C \in [\underline{C}, \overline{C}],\ \ D \in [\underline{D}, \overline{D}]. \tag{4.2}$$

X. Chen, *Analysis and Synthesis of Positive Systems Under ℓ_1 and L_1 Performance*, Springer Theses, DOI 10.1007/978-981-10-2227-2_4

It is noted that the information of the transient state cannot be obtained by designing conventional observers, since they only give an estimate of the state in an asymptotic way. To design a observer which can be used to estimate the state at all times, a lower-bounding estimate $\check{x}(t)$ and an upper-bounding one $\hat{x}(t)$ are introduced. With the two estimates, the signal $x(t)$ can be encapsulated at all times. A pair of observers is proposed as follows:

$$\dot{\check{x}}(t) = \check{F}\check{x}(t) + \check{G}y(t) + \check{K}w(t), \tag{4.3}$$

and

$$\dot{\hat{x}}(t) = \hat{F}\hat{x}(t) + \hat{G}y(t) + \hat{K}w(t), \tag{4.4}$$

where $\hat{x}(t) \in \mathbb{R}^n, \check{x}(t) \in \mathbb{R}^n$. $\check{F}$, $\check{G}$, $\check{K}$, $\hat{F}$, $\hat{G}$ and $\hat{K}$ are observer parameters to be determined.

First, we consider the lower-bounding case. Define the error state $\check{e}(t) = x(t) - \check{x}(t)$; then it follows from systems (4.1) and (4.3) that

$$\dot{\check{e}}(t) = (A - \check{F} - \check{G}C)x(t) + \check{F}\check{e}(t) + (B - \check{G}D - \check{K})w(t). \tag{4.5}$$

Suppose that

$$\check{z}(t) = L\check{e}(t) \tag{4.6}$$

stands for the output of error states and here $L \geq\geq 0$ is known. Now, by defining

$$\begin{aligned} \check{\xi}(t) &= \begin{bmatrix} x(t) \\ \check{e}(t) \end{bmatrix}, \quad \check{A}_\xi = \begin{bmatrix} A & 0 \\ A - \check{F} - \check{G}C & \check{F} \end{bmatrix}, \\ \check{B}_\xi &= \begin{bmatrix} B \\ B - \check{G}D - \check{K} \end{bmatrix}, \quad \check{C}_\xi = \begin{bmatrix} 0 & L \end{bmatrix}, \end{aligned} \tag{4.7}$$

and with (4.1), (4.5) and (4.6), the augmented system is obtained as follows:

$$\begin{cases} \dot{\check{\xi}}(t) = \check{A}_\xi \check{\xi}(t) + \check{B}_\xi w(t), \\ \check{z}(t) = \check{C}_\xi \check{\xi}(t). \end{cases} \tag{4.8}$$

The observer in (4.3) is designed for the positive system in (4.1) to approximate $x(t)$ by $\check{x}(t)$. Consequently, the estimate $\check{x}(t)$ is required to be positive, like system state $x(t)$ itself, which implies that the observer in (4.3) should be a positive system. From Lemma 1.1, we see that $\check{F}$ is Metzler, $\check{G} \geq\geq 0$ and $\check{K} \geq\geq 0$ are needed. The positive lower-bounding observer problem is established as follows.

Positive Lower-Bounding Observer Design (PLOD): *Given a positive system* (4.1) *with* $A \in [\underline{A}, \overline{A}]$, $B \in [\underline{B}, \overline{B}]$, $C \in [\underline{C}, \overline{C}]$, *and* $D \in [\underline{D}, \overline{D}]$, *design a positive observer of the form* (4.3) *with* $\check{F}$ *being Metzler,* $\check{G} \geq\geq 0$ *and* $\check{K} \geq\geq 0$ *such that* *the augmented system* (4.8) *is positive, robustly stable and satisfies the performance* $\|\check{z}\|_{L_1} < \gamma \|w\|_{L_1}$ *under zero initial conditions.*

Similarly, one may define $\hat{e}(t) = \hat{x}(t) - x(t)$ and $\hat{\xi}(t) = [x^T(t),\ \hat{e}^T(t)]^T$. Suppose that

$$\hat{z}(t) = L\hat{e}(t) \tag{4.9}$$

stands for the output of error states. Then we have the augmented system

$$\begin{cases} \dot{\hat{\xi}}(t) = \hat{A}_\xi \hat{\xi}(t) + \hat{B}_\xi w(t), \\ \hat{z}(t) = \hat{C}_\xi \hat{\xi}(t), \end{cases} \tag{4.10}$$

where

$$\hat{A}_\xi = \begin{bmatrix} A & 0 \\ \hat{F} + \hat{G}C - A & \hat{F} \end{bmatrix},$$
$$\hat{B}_\xi = \begin{bmatrix} B \\ \hat{G}D + \hat{K} - B \end{bmatrix},\ \hat{C}_\xi = \begin{bmatrix} 0 & L \end{bmatrix}.$$

In the following, the positive upper-bounding observer design (PUOD) problem is formulated.

Positive Upper-Bounding Observer Design (PUOD): *Given a positive system* (4.1) *with* $A \in [\underline{A}, \overline{A}]$, $B \in [\underline{B}, \overline{B}]$, $C \in [\underline{C}, \overline{C}]$, *and* $D \in [\underline{D}, \overline{D}]$, *design a positive observer of the form* (4.4) *with* $\hat{F}$ *being Metzler,* $\hat{G} \geq\geq 0$ *and* $\hat{K} \geq\geq 0$ *such that the augmented system* (4.10) *is positive, robustly stable and satisfies the performance* $\|\hat{z}\|_{L_1} < \gamma \|w\|_{L_1}$ *under zero initial conditions.*

4.2 Observer Design for Positive Systems

In this section, we propose a pair of positive state-bounding observers which bound the state $x(t)$ at all times, and satisfies the L_1-induced performance. To achieve this, the performance characterization result is first established for the lower-bounding augmented system (4.8). Then, necessary and sufficient conditions are presented for the design of lower-bounding observer. Finally, parallel results are obtained for upper-bounding case.

Based on Theorem 3.3, the following result is derived to serve as a characterization on the stability of lower-bounding augmented system (4.8) with the performance $\|\check{z}\|_{L_1} < \gamma \|w\|_{L_1}$. The proof is omitted here.

Theorem 4.1 *The lower-bounding augmented system in (4.8) is positive, robustly stable and satisfies* $\|\check{z}\|_{L_1} < \gamma \|w\|_{L_1}$ *for any* $A \in [\underline{A}, \overline{A}]$, $B \in [\underline{B}, \overline{B}]$, $C \in [\underline{C}, \overline{C}]$ *and* $D \in [\underline{D}, \overline{D}]$ *under zero initial conditions if and only if there exist Metzler matrix* $\underline{\check{A}}_\xi$, $\underline{\check{B}}_\xi \geq\geq 0$, $\check{C}_\xi \geq\geq 0$ *and a vector* $p \geq\geq 0$ *satisfying*

$$\mathbf{1}^T \check{C}_\xi + p^T \overline{\check{A}}_\xi \ll 0, \tag{4.11}$$

$$p^T \overline{\check{B}}_\xi - \gamma \mathbf{1}^T \ll 0, \tag{4.12}$$

where

$$\underline{\check{A}}_\xi = \begin{bmatrix} \underline{A} & 0 \\ \underline{A} - \check{F} - \check{G}\overline{C} & \check{F} \end{bmatrix}, \ \underline{\check{B}}_\xi = \begin{bmatrix} \underline{B} \\ \underline{B} - \check{G}\overline{D} - \check{K} \end{bmatrix},$$

$$\overline{\check{A}}_\xi = \begin{bmatrix} \overline{A} & 0 \\ \overline{A} - \check{F} - \check{G}\underline{C} & \check{F} \end{bmatrix}, \ \overline{\check{B}}_\xi = \begin{bmatrix} \overline{B} \\ \overline{B} - \check{G}\underline{D} - \check{K} \end{bmatrix}.$$

Then, a necessary and sufficient condition is further established for the existence of the lower-bounding observer.

Theorem 4.2 *Given a stable continuous-time positive system (4.1), a lower-bounding observer (4.3) exists such that the augmented system (4.8) is positive, robustly stable and satisfies* $\|\check{z}\|_{L_1} < \gamma \|w\|_{L_1}$ *for any* $A \in [\underline{A}, \overline{A}]$, $B \in [\underline{B}, \overline{B}]$, $C \in [\underline{C}, \overline{C}]$ *and* $D \in [\underline{D}, \overline{D}]$ *under zero initial conditions if and only if there exist Metzler matrix* $\check{M}_F$, $\check{M}_G \geq\geq 0$, $\check{M}_K \geq\geq 0$ *and vectors* $p_1 \geq\geq 0$, $p_2 \geq\geq 0$ *satisfying*

$$\left[\check{M}_F\right]_{ij} \geq 0, \ i, j = 1, \ldots, n, \ i \neq j, \tag{4.13}$$

$$\left[\check{M}_G\right]_{il} \geq 0, \ i = 1, \ldots, n, \ l = 1, \ldots, p, \tag{4.14}$$

$$\left[\check{M}_K\right]_{ik} \geq 0, \ i = 1, \ldots, n, \ k = 1, \ldots, m, \tag{4.15}$$

$$p_{2i}^T \left[\underline{A}\right]_{ij} - \left[\check{M}_G\right]_{r,i} \left[\overline{C}\right]_{c,j} - \left[\check{M}_F\right]_{ij} \geq 0, \tag{4.16}$$

$$p_{2i}^T \left[\underline{B}\right]_{ik} - \left[\check{M}_G\right]_{r,i} \left[\overline{D}\right]_{c,k} - \left[\check{M}_K\right]_{ik} \geq 0, \tag{4.17}$$

$$p_1^T \overline{A} + p_2^T \overline{A} - \sum_{i=1}^{n} \left[\check{M}_F\right]_{r,i} - \sum_{i=1}^{p} \left[\check{M}_G\right]_{r,i} \underline{C} \ll 0, \tag{4.18}$$

$$\sum_{i=1}^{n} \left[\check{M}_F\right]_{r,i} + \mathbf{1}^T L \ll 0, \tag{4.19}$$

$$p_1^T \overline{B} + p_2^T \overline{B} - \sum_{i=1}^{p} \left[\check{M}_G\right]_{r,i} \underline{D} - \sum_{i=1}^{m} \left[\check{M}_K\right]_{r,i} - \gamma \mathbf{1}^T \ll 0. \tag{4.20}$$

Moreover, a suitable set of $\check{F}$, $\check{G}$ and $\check{K}$ is given by

$$\begin{aligned}\left[\check{F}\right]_{ij} &= p_{2i}^{-1}\left[\check{M}_F\right]_{ij},\\ \left[\check{G}\right]_{il} &= p_{2i}^{-1}\left[\check{M}_G\right]_{il},\\ \left[\check{K}\right]_{ik} &= p_{2i}^{-1}\left[\check{M}_K\right]_{ik}.\end{aligned} \tag{4.21}$$

Proof (*Sufficiency*) Note that $p_2 \geq\geq 0$, it follows from (4.13)–(4.15) and (4.21) that $\check{F}$ is Metzler, $\check{G} \geq\geq 0$ and $\check{K} \geq\geq 0$, which implies that the lower-bounding observer (4.3) is positive.

From (4.21) and $p_2 \geq\geq 0$, (4.16) and (4.17) become

$$\left[\underline{A}\right]_{ij} - \left[\check{G}\right]_{r,i}\left[\overline{C}\right]_{c,j} - \left[\check{F}\right]_{ij} \geq 0, \tag{4.22}$$

$$\left[\underline{B}\right]_{ik} - \left[\check{G}\right]_{r,i}\left[\overline{D}\right]_{c,k} - \left[\check{K}\right]_{ik} \geq 0, \tag{4.23}$$

and we have

$$\underline{A} - \check{G}\overline{C} - \check{F} \geq\geq 0,\ \underline{B} - \check{G}\overline{D} - \check{K} \geq\geq 0. \tag{4.24}$$

Combining (4.24) with $\check{G} \geq\geq 0$ yields the following: for any $A \in [\underline{A}, \overline{A}]$, $B \in [\underline{B}, \overline{B}]$, $C \in [\underline{C}, \overline{C}]$ and $D \in [\underline{D}, \overline{D}]$,

$$\begin{aligned}A - \check{G}C - \check{F} &\geq\geq \underline{A} - \check{G}\overline{C} - \check{F} \geq\geq 0,\\ B - \check{G}D - \check{K} &\geq\geq \underline{B} - \check{G}\overline{D} - \check{K} \geq\geq 0.\end{aligned}$$

Together with $\check{F}$ being Metzler and $L \geq\geq 0$, from (4.7), it shows that the augmented system (4.8) is positive.

From (4.21), we have

$$\begin{aligned}&\sum_{i=1}^{n}\left[\check{M}_F\right]_{r,i} = p_2^T\check{F},\\ &\sum_{i=1}^{p}\left[\check{M}_G\right]_{r,i} = p_2^T\check{G},\ \sum_{i=1}^{m}\left[\check{M}_K\right]_{r,i} = p_2^T\check{K}.\end{aligned} \tag{4.25}$$

With (4.25), inequalities (4.18)–(4.20) equal to

$$\begin{aligned}p_1^T\overline{A} + p_2^T\overline{A} - p_2^T\check{F} - p_2^T\check{G}\underline{C} &\ll 0,\\ p_2^T\check{F} + \mathbf{1}^T L &\ll 0,\\ p_1^T\overline{B} + p_2^T\overline{B} - p_2^T\check{G}\underline{D} - p_2^T\check{K} - \gamma\mathbf{1}^T &\ll 0,\end{aligned}$$

which further imply that

$$\mathbf{1}^T \begin{bmatrix} 0 \; L \end{bmatrix} + p^T \begin{bmatrix} \overline{A} & 0 \\ \overline{A} - \check{F} - \check{G}\underline{C} & \check{F} \end{bmatrix} \ll 0, \tag{4.26}$$

$$p^T \begin{bmatrix} \overline{B} \\ \overline{B} - \check{G}\underline{D} - \check{K} \end{bmatrix} - \gamma \mathbf{1}^T \ll 0, \tag{4.27}$$

where $p^T = \begin{bmatrix} p_1^T \; p_2^T \end{bmatrix}$.

Therefore, by Theorem 4.1, we have that the lower-bounding augmented system (4.8) is robustly stable and satisfies $\|\check{z}\|_{L_1} < \gamma \|w\|_{L_1}$. The sufficiency is proved.

(*Necessity*) Assume that the augmented system (4.8) is robustly stable and satisfies $\|\check{z}\|_{L_1} < \gamma \|w\|_{L_1}$. Then, according to Theorem 4.1, the inequalities (4.26) and (4.27) hold. Denote $p^T \triangleq \begin{bmatrix} p_1^T \; p_2^T \end{bmatrix}$ and we have that the following inequalities hold

$$\begin{aligned} p_1^T \overline{A} + p_2^T \overline{A} - p_2^T \check{F} - p_2^T \check{G}\underline{C} &\ll 0, \\ p_2^T \check{F} + \mathbf{1}^T L &\ll 0, \\ p_1^T \overline{B} + p_2^T \overline{B} - p_2^T \check{G}\underline{D} - p_2^T \check{K} - \gamma \mathbf{1}^T &\ll 0, \end{aligned}$$

Noting that

$$\begin{aligned} p_2^T \check{F} &= \sum_{i=1}^{n} p_{2i} \left[\check{F} \right]_{r,i}, \\ p_2^T \check{G} &= \sum_{i=1}^{p} p_{2i} \left[\check{G} \right]_{r,i}, \; p_2^T \check{K} = \sum_{i=1}^{m} p_{2i} \left[\check{K} \right]_{r,i}, \end{aligned}$$

it turns out that the change of variables

$$\begin{aligned} \left[\check{M}_F \right]_{ij} &= p_{2i} \left[\check{F} \right]_{ij}, \\ \left[\check{M}_G \right]_{il} &= p_{2i} \left[\check{G} \right]_{il}, \; \left[\check{M}_K \right]_{ik} = p_{2i} \left[\check{K} \right]_{ik} \end{aligned}$$

linearizes the problem and yields (4.18)–(4.21).

Since the lower-bounding observer (4.3) is positive, we have $\check{F}$ is Metzler, $\check{G} \geq\geq 0$ and $\check{K} \geq\geq 0$. With $p_2 \geq\geq 0$, the conditions $\check{F}$ is Metzler, $\check{G} \geq\geq 0$ and $\check{K} \geq\geq 0$ equal to (4.13)–(4.15).

In addition, if the lower-bounding augmented system (4.8) is positive, we have

$$A - \check{G}C - \check{F} \geq\geq 0, \; B - \check{G}D - \check{K} \geq\geq 0,$$

which, together with $p_2 \geq\geq 0$, implies (4.16) and (4.17). This completes the whole proof. □

For the performance analysis in the upper-bounding case, the parallel result is presented as follows.

Theorem 4.3 *The upper-bounding augmented system in (4.10) is positive, robustly stable and satisfies* $\|\hat{z}\|_{L_1} < \gamma \|w\|_{L_1}$ *for any* $A \in [\underline{A}, \overline{A}]$, $B \in [\underline{B}, \overline{B}]$, $C \in [\underline{C}, \overline{C}]$ *and* $D \in [\underline{D}, \overline{D}]$ *under zero initial conditions if and only if there exist Metzler matrix* $\underline{\hat{A}}_\xi$, $\underline{\hat{B}}_\xi \geq\geq 0$, $\hat{C}_\xi \geq\geq 0$ *and a vector* $p \geq\geq 0$ *satisfying*

$$\mathbf{1}^T \hat{C}_\xi + p^T \overline{\hat{A}}_\xi \ll 0, \tag{4.28}$$

$$p^T \overline{\hat{B}}_\xi - \gamma \mathbf{1}^T \ll 0, \tag{4.29}$$

where

$$\underline{\hat{A}}_\xi = \begin{bmatrix} \underline{A} & 0 \\ \hat{F} + \hat{G}\underline{C} - \overline{A} & \hat{F} \end{bmatrix}, \quad \underline{\hat{B}}_\xi = \begin{bmatrix} \underline{B} \\ \hat{G}\underline{D} + \hat{K} - \overline{B} \end{bmatrix},$$

$$\overline{\hat{A}}_\xi = \begin{bmatrix} \overline{A} & 0 \\ \hat{F} + \hat{G}\overline{C} - \underline{A} & \hat{F} \end{bmatrix}, \quad \overline{\hat{B}}_\xi = \begin{bmatrix} \overline{B} \\ \hat{G}\overline{D} + \hat{K} - \underline{B} \end{bmatrix}.$$

Next, a necessary and sufficient condition is further established for the existence of the upper-bounding observer. The proof is omitted here.

Theorem 4.4 *Given a stable continuous-time positive system (4.1), an upper-bounding observer (4.4) exists such that the augmented system (4.10) is positive, robustly stable and satisfies* $\|\hat{z}\|_{L_1} < \gamma \|w\|_{L_1}$ *for any* $A \in [\underline{A}, \overline{A}]$, $B \in [\underline{B}, \overline{B}]$, $C \in [\underline{C}, \overline{C}]$ *and* $D \in [\underline{D}, \overline{D}]$ *under zero initial conditions if and only if there exist vectors* $p_1 \geq\geq 0$, $p_2 \geq\geq 0$ *and matrices* $\hat{M}_F$, $\hat{M}_G$, $\hat{M}_K$ *satisfying*

$$\begin{aligned}
&\left[\hat{M}_F\right]_{ij} \geq 0, \ i, j = 1, \ldots, n, \ i \neq j, \\
&\left[\hat{M}_G\right]_{il} \geq 0, \ i = 1, \ldots, n, \ l = 1, \ldots, p, \\
&\left[\hat{M}_K\right]_{ik} \geq 0, \ i = 1, \ldots, n, \ k = 1, \ldots, m, \\
&\left[\hat{M}_F\right]_{ij} + \left[\hat{M}_G\right]_{r,i} \left[\underline{C}\right]_{c,j} - p_{2i}^T \left[\overline{A}\right]_{ij} \geq 0, \\
&\left[\hat{M}_K\right]_{ik} + \left[\hat{M}_G\right]_{r,i} \left[\overline{D}\right]_{c,k} - p_{2i}^T \left[\underline{B}\right]_{ik} \geq 0,
\end{aligned}$$

$$p_1^T \overline{A} + \sum_{i=1}^{n} \left[\hat{M}_F\right]_{r,i} + \sum_{i=1}^{p} \left[\hat{M}_G\right]_{r,i} \overline{C} - p_2^T \underline{A} \ll 0,$$

$$\sum_{i=1}^{n} \left[\hat{M}_F\right]_{r,i} + \mathbf{1}^T L \ll 0,$$

$$p_1^T \overline{B} + \sum_{i=1}^{p} \left[\hat{M}_G\right]_{r,i} \overline{D} + \sum_{i=1}^{m} \left[\hat{M}_K\right]_{r,i} - p_2^T \underline{B} - \gamma \mathbf{1}^T \ll 0.$$

Moreover, a suitable set of $\hat{F}$, $\hat{G}$ and $\hat{K}$ is given by

$$\begin{aligned}
\left[\hat{F}\right]_{ij} &= p_{2i}^{-1}\left[\hat{M}_F\right]_{ij}, \\
\left[\hat{G}\right]_{il} &= p_{2i}^{-1}\left[\hat{M}_G\right]_{il}, \\
\left[\hat{K}\right]_{ik} &= p_{2i}^{-1}\left[\hat{M}_K\right]_{ik}.
\end{aligned}$$

4.3 Numerical Example

An illustrative example is presented in this section to illustrate the effectiveness of the theoretical results.

Consider system

$$\begin{cases} \dot{x}(t) = Ax(t) + Bw(t), \\ y(t) = Cx(t) + Dw(t), \end{cases} \tag{4.30}$$

with

$$\begin{aligned}
A &= \begin{bmatrix} -1.8 \pm 0.03 & 0.1 & 0.1 \pm 0.03 \\ 0.2 & -1.6 \pm 0.01 & 0.6 \\ 0.5 & 0.2 & -1.4 \pm 0.03 \end{bmatrix}, \\
B &= \begin{bmatrix} 0.2 \pm 0.02 \\ 0.5 \\ 0.1 \pm 0.03 \end{bmatrix}, \\
C &= \begin{bmatrix} 0.1 & 0.5 \pm 0.01 & 0.2 \end{bmatrix}, \quad D = 0.1 \pm 0.03.
\end{aligned}$$

Here, assume $L = \begin{bmatrix} 0.2 & 0.5 & 0.2 \end{bmatrix}$ and $\gamma = 0.15$. By solving the conditions in Theorem 4.2 via Yalmip, a feasible solution is obtained as follows:

$$\begin{aligned}
p_1 &= \begin{bmatrix} 0.0937 & 0.0470 & 0.0223 \end{bmatrix}^T, \\
p_2 &= \begin{bmatrix} 0.4424 & 0.4350 & 0.9906 \end{bmatrix}^T,
\end{aligned}$$

which further yields the matrices of the lower-bounding observer as

$$\begin{aligned}
\check{F} &= \begin{bmatrix} -1.9063 & 0.0561 & 0.0300 \\ 0.1361 & -1.7054 & 0.2606 \\ 0.4592 & 0.1570 & -1.1489 \end{bmatrix}, \\
\check{G} &= \begin{bmatrix} 0.0454 \\ 0.1272 \\ 0.0567 \end{bmatrix}, \quad \check{K} = \begin{bmatrix} 0.1641 \\ 0.4730 \\ 0.0582 \end{bmatrix}.
\end{aligned}$$

Fig. 4.1 State x(t) and its state-bounding estimate. [2014] IEEE. Reprinted, with permission, from Ref. [1]

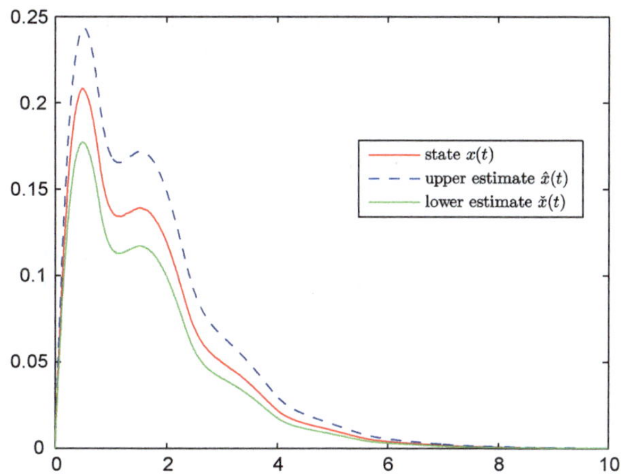

Similarly, by solving the conditions in Theorem 4.4, with $\gamma = 0.15$, a feasible solution is achieved with

$$p_1 = \begin{bmatrix} 0.0486 \ 0.0870 \ 0.0930 \end{bmatrix}^T,$$
$$p_2 = \begin{bmatrix} 0.2825 \ 0.1911 \ 0.7894 \end{bmatrix}^T,$$

which further yields the matrices of the upper-bounding observer as

$$\hat{F} = \begin{bmatrix} -1.7845 & 0.0074 & 0.0941 \\ 0.2023 & -1.0645 & 0.6040 \\ 0.4608 & 0.0006 & -1.4504 \end{bmatrix},$$
$$\hat{G} = \begin{bmatrix} 0.1940 \\ 0.0074 \\ 0.4091 \end{bmatrix}, \quad \hat{K} = \begin{bmatrix} 0.2084 \\ 0.5025 \\ 0.1021 \end{bmatrix}.$$

With input $w(t) = 4.5e^{-t}|\cos(2t)|$ and zero initial conditions, Fig. 4.1 depicts the state $x(t)$, the lower estimate $\check{x}(t)$ and the upper estimate $\hat{x}(t)$.

4.4 Summary

In this chapter, the problem of positive observers for interval positive systems with L_1-induced performance has been studied. A new characterization on the L_1-induced performance of the augmented system has been established. Based on the novel performance characterization, conditions have been derived for the existence of state-bounding positivity preserving observers. Moreover, the designed observers can provide an estimate of the state in an asymptotic way. In addition, all the conditions have been given under the LP framework and thus can be easily verified. Finally, an example has been proposed to demonstrate the effectiveness of the proposed approach.

Reference

1. Chen X, Lam J (2014) Positive state-bounding observer for interval positive systems under L_1 performance. In: Proceedings of the 33rd Chinese Control Conference

Chapter 5
Positive Filtering for Positive Systems Under L_1 Performance

In this chapter, the positive filtering problem is addressed for positive continuous-time systems under the L_1-induced performance. A pair of positive filters with error-bounding feature is proposed to estimate the output of positive systems. A novel characterization is first obtained to ensure the filtering error system asymptotically stable with a prescribed L_1-induced performance. Then, necessary and sufficient conditions for the existence of required filters are presented, and the obtained results are expressed in terms of linear programming problems, which can be easily checked by standard software. Finally, a numerical example is given to illustrate the effectiveness of the proposed design procedures.

This chapter is briefly organized as follows. The positivity preserving error-bounding filtering problem for positive systems is formulated in Sect. 5.1. Section 5.2 is devoted to the positive error-bounding filter synthesis for positive systems. An example is provided in Sect. 5.3 to show the application of the theoretical results. Finally, Sect. 5.4 gives a summary of this chapter.

5.1 Problem Formulation

Consider the following stable system:

$$\begin{cases} \dot{x}(t) = Ax(t) + Bw(t), \\ y(t) = Cx(t) + Dw(t), \\ z(t) = Lx(t), \end{cases} \tag{5.1}$$

where $x(t) \in \mathbb{R}^n$ is the state vector, $w(t) \in \mathbb{R}^m$ is the disturbance signal, $y(t) \in \mathbb{R}^r$ is the measurement, $z(t) \in \mathbb{R}^q$ is the signal to be determined; A, B, C, D and L are system matrices with compatible dimensions. System (5.1) is positive if for all

X. Chen, *Analysis and Synthesis of Positive Systems Under ℓ_1 and L_1 Performance*, Springer Theses, DOI 10.1007/978-981-10-2227-2_5

$x(0) \geq\geq 0$ and all input $w(t) \geq\geq 0$, we have $x(t) \geq\geq 0$, $y(t) \geq\geq 0$ and $z(t) \geq\geq 0$ for $t > 0$.

It is noted that conventional filters only provide an estimate of the system output in an asymptotic way, which further implies that no information about the transient output can be given under this framework. In this chapter, a filter which can be used to estimate the output at all times is designed. To achieve this, two nonnegative estimates $\check{z}(t)$ and $\hat{z}(t)$, which encapsulate the signal $z(t)$ for all times, are introduced. Different from previous work, a pair of filters is proposed with a general structure as follows:

$$\begin{cases} \dot{\check{x}}(t) = \underline{A}_f \check{x}(t) + \underline{B}_f y(t), \\ \check{z}(t) = \underline{C}_f \check{x}(t), \end{cases} \tag{5.2}$$

and

$$\begin{cases} \dot{\hat{x}}(t) = \overline{A}_f \hat{x}(t) + \overline{B}_f y(t), \\ \hat{z}(t) = \overline{C}_f \hat{x}(t), \end{cases} \tag{5.3}$$

where $\hat{x}(t) \in \mathbb{R}^n$, $\check{x}(t) \in \mathbb{R}^n$, $\hat{z}(t) \in \mathbb{R}^q$ and $\check{z}(t) \in \mathbb{R}^q$. $\underline{A}_f$, $\overline{A}_f$, $\underline{B}_f$, $\overline{B}_f$, $\underline{C}_f$ and $\overline{C}_f$ are filtering parameters to be determined later.

In the following, the lower-bounding case is first considered.

By defining new variables $\check{x}_e(t) = x(t) - \check{x}(t)$, $\check{\xi}(t) = [x^T(t),\ \check{x}_e^T(t)]^T$ and $\check{e}(t) = z(t) - \check{z}(t)$, it follows from systems (5.1) and (5.2) that the augmented system can be described by

$$\begin{cases} \dot{\check{\xi}}(t) = \underline{A}_\xi \check{\xi}(t) + \underline{B}_\xi w(t), \\ \check{e}(t) = \underline{C}_\xi \check{\xi}(t), \end{cases} \tag{5.4}$$

where

$$\begin{aligned} \underline{A}_\xi &= \begin{bmatrix} A & 0 \\ A - \underline{B}_f C - \underline{A}_f & \underline{A}_f \end{bmatrix}, \quad \underline{B}_\xi = \begin{bmatrix} B \\ B - \underline{B}_f D \end{bmatrix}, \\ \underline{C}_\xi &= \begin{bmatrix} L - \underline{C}_f & \underline{C}_f \end{bmatrix}. \end{aligned} \tag{5.5}$$

The purpose of the filter (5.2) for the positive system (5.1) is to approximate $z(t)$ by $\check{z}(t)$. Therefore, the filter should be chosen in such a way that the estimator $\check{z}(t)$ should be positive, like $z(t)$ itself. To ensure the positivity of $\check{z}(t)$, it is natural to require that the filter (5.2) should be a positive system. To this end, according to Lemma 1.1, it is natural to require that $\underline{A}_f$ is Metzler, $\underline{B}_f \geq\geq 0$ and $\underline{C}_f \geq\geq 0$. Thus, the positive lower-bounding filtering (PLF) problem can be formulated as follows.

Positive Lower-bounding Filtering (PLF): *Given a stable positive system (5.1), design a positive filter of the form (5.2) with $\underline{A}_f$ being Metzler, $\underline{B}_f \geq\geq 0$ and $\underline{C}_f \geq\geq 0$ such that the filtering error system (5.4) is positive, asymptotically stable for $w(t) = 0$ and satisfies the performance $\|\check{e}\|_{L_1} < \gamma_l \|w\|_{L_1}$ under zero initial conditions.*

Similarly, one may define $\hat{x}_e(t) = \hat{x}(t) - x(t)$, $\hat{\xi}(t) = [x^T(t),\ \hat{x}_e^T(t)]^T$ and $\hat{e}(t) = \hat{z}(t) - z(t)$, the augmented system can be described by

$$\begin{cases} \dot{\hat{\xi}}(t) = \overline{A}_\xi \hat{\xi}(t) + \overline{B}_\xi w(t), \\ \hat{e}(t) = \overline{C}_\xi \hat{\xi}(t), \end{cases} \tag{5.6}$$

where

$$\overline{A}_\xi = \begin{bmatrix} A & 0 \\ \overline{A}_f + \overline{B}_f C - A & \overline{A}_f \end{bmatrix},\ \overline{B}_\xi = \begin{bmatrix} B \\ \overline{B}_f D - B \end{bmatrix},\ \overline{C}_\xi = \begin{bmatrix} \overline{C}_f - L & \overline{C}_f \end{bmatrix}.$$

In the following, the corresponding positive upper-bounding filtering (PUF) problem is formulated in a similar way.

Positive Upper-bounding Filtering (PUF): *Given a stable positive system (5.1), design a positive filter of the form (5.3) with $\overline{A}_f$ being Metzler, $\overline{B}_f \geq\geq 0$ and $\overline{C}_f \geq\geq 0$ such that the filtering error system (5.6) is positive, asymptotically stable for $w(t) = 0$ and satisfies the performance $\|\hat{e}\|_{L_1} < \gamma_u \|w\|_{L_1}$ under zero initial conditions.*

Remark 5.1 In this chapter, the error signal is supposed to be nonnegative. This specification facilitates the synthesis of the desired positive filters. As for removing the constraint on $\check{e}(t)$ and $\hat{e}(t)$, it remains unsolved and needs further study.

5.2 Filter Design for Positive Systems

In this section, we design a pair of positive error-bounding filters which bound the signal $z(t)$ at all times, and satisfy the L_1-induced performance. To achieve this, the performance characterization result is first established for the lower-bounding augmented system (5.4). Then, necessary and sufficient conditions are presented for the design of a lower-bounding filter, which are expressed in terms of linear programming problems. Then, parallel results are obtained for the upper-bounding case. Finally, a method to design the L_1-induced error-bounding filters is presented by combining the two separate cases.

First, the following result is derived which provides a fundamental characterization on the stability of lower-bounding filtering error system (5.4) with the performance $\|\check{e}\|_{L_1} < \gamma_l \|w\|_{L_1}$. Based on Lemma 3.2, the performance characterization of lower-bounding filtering error system (5.4) can be directly obtained and thus the detailed proof is omitted here.

Theorem 5.1 *Suppose the filtering error system in (5.4) is positive. System (5.4) is asymptotically stable and satisfies $\|\check{e}\|_{L_1} < \gamma_l \|w\|_{L_1}$ if and only if there exists a vector $p_l \geq\geq 0$ satisfying*

$$\mathbf{1}^T \underline{C}_{\xi} + p_l^T \underline{A}_{\xi} \ll 0, \tag{5.7}$$
$$p_l^T \underline{B}_{\xi} - \gamma_l \mathbf{1}^T \ll 0. \tag{5.8}$$

Then, a necessary and sufficient condition is further established for the existence of the lower-bounding filter.

Theorem 5.2 *Given a stable continuous-time positive system (5.1), a lower-bounding positive filter (5.2) exists such that the filtering error system (5.4) is positive, asymptotically stable and satisfies* $\|\breve{e}\|_{L_1} < \gamma_l \|w\|_{L_1}$ *if and only if there exist vectors* $p_{l1} \geq\geq 0$, $p_{l2} \geq\geq 0$ *and Metzler matrix* $\underline{M}_{A_f}$, $\underline{M}_{B_f} \geq\geq 0$, $\underline{C}_f \geq\geq 0$ *satisfying*

$$L - \underline{C}_f \geq\geq 0, \tag{5.9}$$
$$p_{l2i}^T [A]_{ij} - [\underline{M}_{B_f}]_{r,i} [C]_{c,j} - [\underline{M}_{A_f}]_{ij} \geq 0, \tag{5.10}$$
$$p_{l2i}^T [B]_{ik} - [\underline{M}_{B_f}]_{r,i} [D]_{c,k} \geq 0, \tag{5.11}$$
$$\mathbf{1}^T (L - \underline{C}_f) + p_{l1}^T A + p_{l2}^T A - \sum_{i=1}^{n} [\underline{M}_{B_f}]_{r,i} C - \sum_{i=1}^{n} [\underline{M}_{A_f}]_{r,i} \ll 0, \tag{5.12}$$
$$\mathbf{1}^T \underline{C}_f + \sum_{i=1}^{n} [\underline{M}_{A_f}]_{r,i} \ll 0, \tag{5.13}$$
$$p_{l1}^T B + p_{l2}^T B - \sum_{i=1}^{n} [\underline{M}_{B_f}]_{r,i} D - \gamma_l \mathbf{1}^T \ll 0. \tag{5.14}$$

Moreover, a suitable set of $\underline{A}_f$ *and* $\underline{B}_f$ *is given by*

$$[\underline{A}_f]_{ij} = p_{l2i}^{-1} [\underline{M}_{A_f}]_{ij}, \quad [\underline{B}_f]_{ik} = p_{l2i}^{-1} [\underline{M}_{B_f}]_{ik}. \tag{5.15}$$

Proof (*Sufficiency*) Note that $p_{l2} \geq\geq 0$, it follows from (5.15) that $\underline{A}_f$ is Metzler and $\underline{B}_f \geq\geq 0$. Together with $\underline{C}_f \geq\geq 0$, it implies that the lower-bounding filter (5.2) is positive.

From (5.15) and $p_{l2} \geq\geq 0$, (5.10) and (5.11) become

$$[A]_{ij} - [\underline{B}_f]_{r,i} [C]_{c,j} - [\underline{A}_f]_{ij} \geq 0, \tag{5.16}$$
$$[B]_{ik} - [\underline{B}_f]_{r,i} [D]_{c,k} \geq 0, \tag{5.17}$$

which indicates that

$$A - \underline{B}_f C - \underline{A}_f \geq\geq 0,$$
$$B - \underline{B}_f D \geq\geq 0.$$

Together with $\underline{A}_f$ being Metzler, $\underline{B}_f \geq\geq 0$, $\underline{C}_f \geq\geq 0$ and (5.9), from (5.5), it shows that the filtering error system (5.4) is positive.

Then, from (5.15), we have

$$\sum_{i=1}^{n} [\underline{M}_{A_f}]_{r,i} = p_{l2}^T \underline{A}_f, \quad \sum_{i=1}^{n} [\underline{M}_{B_f}]_{r,i} = p_{l2}^T \underline{B}_f. \tag{5.18}$$

With (5.18), inequalities (5.12)–(5.14) equal to

$$\begin{aligned} \mathbf{1}^T(L - \underline{C}_f) + p_{l1}^T A + p_{l2}^T(A - \underline{B}_f C - \underline{A}_f) &\ll 0, \\ \mathbf{1}^T \underline{C}_f + p_{l2}^T \underline{A}_f &\ll 0, \\ p_{l1}^T B + p_{l2}^T(B - \underline{B}_f D) - \gamma_l \mathbf{1}^T &\ll 0, \end{aligned}$$

which further imply that

$$\mathbf{1}^T \begin{bmatrix} L - \underline{C}_f & \underline{C}_f \end{bmatrix} + p_l^T \begin{bmatrix} A & 0 \\ A - \underline{B}_f C - \underline{A}_f & \underline{A}_f \end{bmatrix} \ll 0, \tag{5.19}$$

$$p_l^T \begin{bmatrix} B \\ B - \underline{B}_f D \end{bmatrix} - \gamma_l \mathbf{1}^T \ll 0, \tag{5.20}$$

where $p_l^T = \begin{bmatrix} p_{l1}^T & p_{l2}^T \end{bmatrix}$.

Therefore, by Theorem 5.1, we have that the filtering error system (5.4) asymptotically stable and satisfies $\|\check{e}\|_{L_1} < \gamma_l \|w\|_{L_1}$. This proves the sufficiency.

(*Necessity*) Assume that the filtering error system (5.4) is asymptotically stable and satisfies $\|\check{e}\|_{L_1} < \gamma_l \|w\|_{L_1}$. Then, according to Theorem 5.1, the inequalities (5.19) and (5.20) hold. Denote $p_l^T \triangleq \begin{bmatrix} p_{l1}^T & p_{l2}^T \end{bmatrix}$ and we have the following inequalities hold:

$$\mathbf{1}^T(L - \underline{C}_f) + p_{l1}^T A + p_{l2}^T A - p_{l2}^T \underline{B}_f C - p_{l2}^T \underline{A}_f \ll 0, \tag{5.21}$$

$$\mathbf{1}^T \underline{C}_f + p_{l2}^T \underline{A}_f \ll 0, \tag{5.22}$$

$$p_{l1}^T B + p_{l2}^T B - p_{l2}^T \underline{B}_f D - \gamma_l \mathbf{1}^T \ll 0. \tag{5.23}$$

Noting that $p_{l2}^T \underline{A}_f = \sum_{i=1}^{n} p_{l2i} [\underline{A}_f]_{r,i}$ and $p_{l2}^T \underline{B}_f = \sum_{i=1}^{n} p_{l2i} [\underline{B}_f]_{r,i}$, it turns out that the change of variables

$$[\underline{M}_{A_f}]_{ij} = p_{l2i} [\underline{A}_f]_{ij}, \quad [\underline{M}_{B_f}]_{ik} = p_{l2i} [\underline{B}_f]_{ik} \tag{5.24}$$

linearizes the problem and yields (5.12)–(5.15).

Since the lower-bounding filter (5.2) is positive, we have $\underline{A}_f$ is Metzler, $\underline{B}_f \geq\geq 0$ and $\underline{C}_f \geq\geq 0$. With (5.24) and $p_{l2} \geq\geq 0$, the conditions $\underline{A}_f$ is Metzler and $\underline{B}_f \geq\geq 0$ equal to $\underline{M}_{A_f}$ is Metzler and $\underline{M}_{B_f} \geq\geq 0$.

In addition, if the filtering error system (5.4) is positive, we have

$$L - \underline{C}_f \geq\geq 0, \ [A]_{ij} - [\underline{B}_f]_{r,i}[C]_{c,j} - [\underline{A}_f]_{ij} \geq 0, \tag{5.25}$$
$$[B]_{ik} - [\underline{B}_f]_{r,i}[D]_{c,k} \geq 0, \tag{5.26}$$

which, together with (5.24) and $p_{l2} \geq\geq 0$, implies (5.9)–(5.11). This completes the whole proof. □

For the performance analysis in the upper-bounding case, similar results can be readily obtained, and we present the parallel result in the following theorem with the proof omitted.

Theorem 5.3 *Suppose the filtering error system in (5.6) is positive. System (5.6) is asymptotically stable and satisfies $\|\hat{e}\|_{L_1} < \gamma_u \|w\|_{L_1}$ if and only if there exists a vector $p_u \geq\geq 0$ satisfying*

$$\mathbf{1}^T \overline{C}_\xi + p_u^T \overline{A}_\xi \ll 0, \tag{5.27}$$
$$p_u^T \overline{B}_\xi - \gamma_u \mathbf{1}^T \ll 0. \tag{5.28}$$

Similarly as for the lower-bounding case, a necessary and sufficient condition is proposed for the existence of the upper-bounding filter. The proof is omitted here.

Theorem 5.4 *Given a stable continuous-time positive system (5.1), an upper-bounding positive filter (5.3) exists such that the the filtering error system (5.6) is positive, asymptotically stable and satisfies $\|\hat{e}\|_{L_1} < \gamma_u \|w\|_{L_1}$ if and only if there exist vectors $p_{u1} \geq\geq 0$, $p_{u2} \geq\geq 0$ and Metzler matrix $\overline{M}_{A_f}$, $\overline{M}_{B_f} \geq\geq 0$, $\overline{C}_f \geq\geq 0$ satisfying*

$$\overline{C}_f - L \geq\geq 0, \tag{5.29}$$
$$[\overline{M}_{A_f}]_{ij} + [\overline{M}_{B_f}]_{r,i}[C]_{c,j} - p_{u2i}^T[A]_{ij} \geq 0, \tag{5.30}$$
$$[\overline{M}_{B_f}]_{r,i}[D]_{c,k} - p_{u2i}^T[B]_{ik} \geq 0, \tag{5.31}$$
$$\mathbf{1}^T(\overline{C}_f - L) + p_{u1}^T A - p_{u2}^T A + \sum_{i=1}^{n}[\overline{M}_{B_f}]_{r,i} C + \sum_{i=1}^{n}[\overline{M}_{A_f}]_{r,i} \ll 0, \tag{5.32}$$
$$\mathbf{1}^T \overline{C}_f + \sum_{i=1}^{n}[\overline{M}_{A_f}]_{r,i} \ll 0, \tag{5.33}$$
$$p_{u1}^T B - p_{u2}^T B + \sum_{i=1}^{n}[\overline{M}_{B_f}]_{r,i} D - \gamma_u \mathbf{1}^T \ll 0. \tag{5.34}$$

Moreover, a suitable set of $\overline{A}_f$ and $\overline{B}_f$ is given by

$$[\overline{A}_f]_{ij} = p_{u2i}^{-1}[\overline{M}_{A_f}]_{ij}, \ [\overline{B}_f]_{ik} = p_{u2i}^{-1}[\overline{M}_{B_f}]_{ik}. \tag{5.35}$$

Problems PLF and PUF have been considered as two separate problems. The conditions obtained in Theorems 5.2 and 5.4 are expressed in terms of linear programming problems, which can be verified easily by effective algorithms. In the following, we will show that the L_1-induced error-bounding filters can be designed as a combined method.

First, we define new variables $\xi(t) = [x^T(t),\ \check{x}_e^T(t),\ \hat{x}_e^T(t)]^T$ and $e(t) = \hat{z}(t) - \check{z}(t)$, the augmented system is described by

$$\begin{cases} \dot{\xi}(t) = A_\xi \xi(t) + B_\xi w(t), \\ e(t) = C_\xi \xi(t), \end{cases} \tag{5.36}$$

where

$$A_\xi = \begin{bmatrix} A & 0 & 0 \\ A - \underline{B}_f C - \underline{A}_f & \underline{A}_f & 0 \\ \overline{A}_f + \overline{B}_f C - A & 0 & \overline{A}_f \end{bmatrix}, \quad B_\xi = \begin{bmatrix} B \\ B - \underline{B}_f D \\ \overline{B}_f D - B \end{bmatrix},$$
$$C_\xi = \begin{bmatrix} \overline{C}_f - \underline{C}_f & \underline{C}_f & \overline{C}_f \end{bmatrix}.$$

Next, the L_1-induced positive error-bounding filtering (PEF) problem can be formulated as follows.

Positive Error-bounding Filtering (PEF): *Given a stable positive system* (5.1)*, design positive error-bounding filters of the form* (5.2) *and* (5.3) *such that the filtering error system* (5.36) *is positive, asymptotically stable for* $w(t) = 0$ *and satisfies the performance* $\|e\|_{L_1} < \gamma\|w\|_{L_1}$ *under zero initial conditions. That is, the following conditions should be satisfied simultaneously*:

1. $\underline{A}_f, \overline{A}_f$ *are Metzler,* $\underline{B}_f$*,* $\underline{C}_f$*,* $\overline{B}_f$ *and* $\overline{C}_f$ *are nonnegative*;
2. *Systems* (5.4) *and* (5.6) *are positive*;
3. *System* (5.36) *is asymptotically stable for* $w(t) = 0$ *and satisfies the performance* $\|e\|_{L_1} < \gamma\|w\|_{L_1}$.

Now, we are in a position to present the result in the following theorem with the proof omitted.

Theorem 5.5 *Given a stable continuous-time positive system (5.1), a lower-bounding positive filter (5.2) and an upper-bounding positive filter (5.3) exist such that system (5.36) is asymptotically stable and satisfies* $\|e\|_{L_1} < \gamma\|w\|_{L_1}$ *if and only if there exist vectors* $p_1 \geq\geq 0$, $p_2 \geq\geq 0$, $p_3 \geq\geq 0$ *and Metzler matrices* $\underline{M}_{A_f}$, $\overline{M}_{A_f}$, $\underline{M}_{B_f} \geq\geq 0$, $\overline{M}_{B_f} \geq\geq 0$, $\underline{C}_f \geq\geq 0$, $\overline{C}_f \geq\geq 0$ *satisfying*

$$L - \underline{C}_f \geq\geq 0, \tag{5.37}$$
$$p_{2i}^T [A]_{ij} - [\underline{M}_{B_f}]_{r,i} [C]_{c,j} - [\underline{M}_{A_f}]_{ij} \succeq 0, \tag{5.38}$$
$$p_{2i}^T [B]_{ik} - [\underline{M}_{B_f}]_{r,i} [D]_{c,k} \succeq 0, \tag{5.39}$$
$$\overline{C}_f - L \geq\geq 0, \tag{5.40}$$

$$[\overline{M}_{A_f}]_{ij} + [\overline{M}_{B_f}]_{r,i}[C]_{c,j} - p_{3i}^T[A]_{ij} \geq\geq 0, \quad (5.41)$$

$$[\overline{M}_{B_f}]_{r,i}[D]_{c,k} - p_{3i}^T[B]_{ik} \geq\geq 0, \quad (5.42)$$

$$\begin{aligned} \mathbf{1}^T(\overline{C}_f - \underline{C}_f) + p_1^T A + p_2^T A - \sum_{i=1}^n [\underline{M}_{B_f}]_{r,i} C - \sum_{i=1}^n [\underline{M}_{A_f}]_{r,i} \\ + \sum_{i=1}^n [\overline{M}_{B_f}]_{r,i} C + \sum_{i=1}^n [\overline{M}_{A_f}]_{r,i} - p_3^T A \ll 0, \quad (5.43) \end{aligned}$$

$$\mathbf{1}^T \underline{C}_f + \sum_{i=1}^n [\underline{M}_{A_f}]_{r,i} \ll 0, \quad (5.44)$$

$$\mathbf{1}^T \overline{C}_f + \sum_{i=1}^n [\overline{M}_{A_f}]_{r,i} \ll 0, \quad (5.45)$$

$$p_1^T B + p_2^T B - \sum_{i=1}^n [\underline{M}_{B_f}]_{r,i} D - p_3^T B + \sum_{i=1}^n [\overline{M}_{B_f}]_{r,i} D - \gamma \mathbf{1}^T \ll 0. \quad (5.46)$$

Moreover, a suitable set of $\underline{A}_f$, $\underline{B}_f$, $\overline{A}_f$ *and* $\overline{B}_f$ *is given by*

$$\begin{aligned} [\underline{A}_f]_{ij} &= p_{2i}^{-1}[\underline{M}_{A_f}]_{ij}, \quad [\underline{B}_f]_{ik} = p_{2i}^{-1}[\underline{M}_{B_f}]_{ik}, \\ [\overline{A}_f]_{ij} &= p_{3i}^{-1}[\overline{M}_{A_f}]_{ij}, \quad [\overline{B}_f]_{ik} = p_{3i}^{-1}[\overline{M}_{B_f}]_{ik}. \end{aligned} \quad (5.47)$$

Remark 5.2 Theorem 5.5 can be seen as a combination of Theorems 5.2 and 5.4. In fact, optimally designing them together through Theorem 5.5 is the same as optimally designing the two filters separately through Theorems 5.2 and 5.4. Note that $e(t) = \hat{e}(t) + \check{e}(t)$, which together with the definition of L_1-induced norm and the conditions $\check{e}(t) \geq\geq 0$, $\hat{e}(t) \geq\geq 0$, yields the following equation:

$$\|e\|_{L_1} = \|\hat{e}\|_{L_1} + \|\check{e}\|_{L_1}.$$

The proofs of Theorems 5.4 and 5.5 are similar to that of Theorem 5.2 and from these proofs, it is not difficult to see that the optimal values of γ, γ_l and γ_u have the relationship that

$$\gamma^* = \gamma_l^* + \gamma_u^*$$

and correspondingly, the feasible solutions satisfy

$$p_1 = p_{u1} + p_{l1}, \; p_2 = p_{l2}, \; p_3 = p_{u2}.$$

5.3 Numerical Example

In this section, we present an illustrative example to demonstrate the effectiveness of the proposed approach.

Consider system (5.1) with the following system matrices:

$$A = \begin{bmatrix} -0.8 & 0.1 & 0.2 \\ 0.2 & -1.1 & 0.1 \\ 0.15 & 0.2 & -0.5 \end{bmatrix}, \quad B = \begin{bmatrix} 0.1 \\ 0.5 \\ 0.4 \end{bmatrix}, \quad C = \begin{bmatrix} 0.8 \; 0.8 \; 1 \end{bmatrix},$$
$$D = 1, \; L = \begin{bmatrix} 0.5 \; 0.3 \; 0.2 \end{bmatrix}.$$

By solving the conditions in Theorem 5.2 via Yalmip, we obtain an upper bound of γ_l^* as 0.1802 and a feasible solution is found to be

$$p_{l1} = \begin{bmatrix} 0.0001 \; 0.0001 \; 0.0000 \end{bmatrix}^T,$$
$$p_{l2} = \begin{bmatrix} 1.7267 \; 0.1624 \; 0.5421 \end{bmatrix}^T,$$

which further yields the matrices of the lower-bounding filter as

$$\underline{A}_f = \begin{bmatrix} -0.8800 & 0.0200 & 0.1000 \\ 0.1200 & -1.1800 & 0.0000 \\ 0.0000 & 0.0500 & -0.6875 \end{bmatrix}, \quad \underline{B}_f = \begin{bmatrix} 0.1000 \\ 0.1000 \\ 0.1875 \end{bmatrix},$$
$$\underline{C}_f = \begin{bmatrix} 1.5000 \; 0.1300 \; 0.2000 \end{bmatrix}.$$

Similarly, by solving the conditions in Theorem 5.4, we obtain an upper bound of γ_u^* as 0.1327 and a feasible solution is found to be

$$p_{u1} = \begin{bmatrix} 0.1705 \; 0.0921 \; 0.1741 \end{bmatrix}^T,$$
$$p_{u2} = \begin{bmatrix} 1.7046 \; 0.1094 \; 0.4116 \end{bmatrix}^T,$$

which further yields the matrices of the upper-bounding filter as

$$\overline{A}_f = \begin{bmatrix} -0.8800 & 0.0200 & 0.1000 \\ 0.0000 & -1.5000 & 0.0000 \\ 0.0000 & 0.0000 & -0.9000 \end{bmatrix}, \quad \overline{B}_f = \begin{bmatrix} 0.1000 \\ 0.5000 \\ 0.4000 \end{bmatrix},$$
$$\overline{C}_f = \begin{bmatrix} 1.5000 \; 0.1300 \; 0.2000 \end{bmatrix}.$$

On the other hand, by solving the conditions in Theorem 5.5, we obtain an upper bound of γ^* as 0.3130 and a feasible solution is found to be

$$p_1 = \begin{bmatrix} 0.1706 \; 0.0921 \; 0.1742 \end{bmatrix}^T,$$
$$p_2 = \begin{bmatrix} 1.7267 \; 0.1624 \; 0.5421 \end{bmatrix}^T,$$
$$p_3 = \begin{bmatrix} 1.7046 \; 0.1094 \; 0.4116 \end{bmatrix}^T,$$

which further yields the matrices of the lower-bounding and upper-bounding filters as

$$\underline{A}_f = \begin{bmatrix} -0.8800 & 0.0200 & 0.1000 \\ 0.1200 & -1.1800 & 0.0000 \\ 0.0000 & 0.0500 & -0.6875 \end{bmatrix}, \; \underline{B}_f = \begin{bmatrix} 0.1000 \\ 0.1000 \\ 0.1875 \end{bmatrix},$$
$$\overline{A}_f = \begin{bmatrix} -0.8800 & 0.0200 & 0.1000 \\ 0.0000 & -1.5000 & 0.0000 \\ 0.0000 & 0.0000 & -0.9000 \end{bmatrix}, \; \overline{B}_f = \begin{bmatrix} 0.1000 \\ 0.5000 \\ 0.4000 \end{bmatrix},$$
$$\underline{C}_f = \begin{bmatrix} 1.5000 \; 0.1300 \; 0.2000 \end{bmatrix}, \overline{C}_f = \begin{bmatrix} 1.5000 \; 0.1300 \; 0.2000 \end{bmatrix}.$$

It is noted that the optimal values of performance level and feasible solutions satisfy the following equations:

$$\gamma^* = \gamma_l^* + \gamma_u^*, \; p_1 = p_{u1} + p_{l1}, \; p_2 = p_{l2}, \; p_3 = p_{u2}.$$

Moreover, it is shown that Theorem 5.5 gives the same optimal filters with the two separate filters designed through Theorems 5.2 and 5.4.

With input $w(t) = 4.5e^{-t}|\cos(2t)|$ and zero initial conditions, Fig. 5.1 depicts the trajectories of output $z(t)$ of the original positive system, the lower estimate $\check{z}(t)$ and the upper estimate $\hat{z}(t)$. Figure 5.2 shows the input $w(t)$.

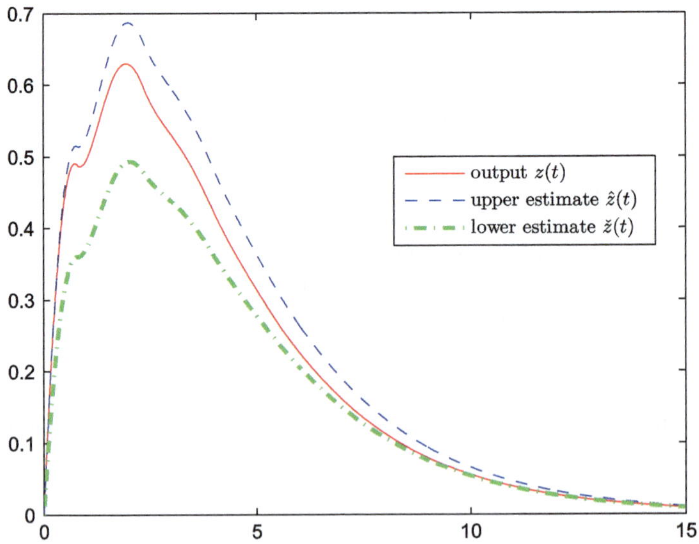

Fig. 5.1 Output $z(t)$ and its estimates. Reprinted from Ref. [1] by permission of Taylor and Francis Ltd

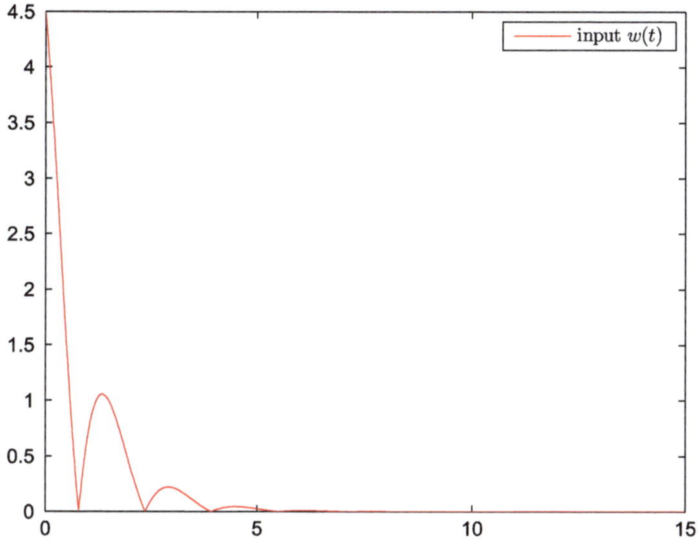

Fig. 5.2 Input $w(t)$. Reprinted from Ref. [1] by permission of Taylor and Francis Ltd

5.4 Summary

In this chapter, the problem of positive filtering for positive systems with L_1-induced performance has been studied. New characterization on the L_1-induced performance of the filtering error system has been established. Necessary and sufficient conditions for the existence of positive error-bounding filters have been developed, where the designed filters can be used to estimate the output signal of the positive systems at all times. It should be pointed out that all the conditions are obtained under the LP framework, which can be easily verified by standard software. Finally, an example has been presented to illustrate the effectiveness of the theoretical results.

Reference

1. Chen X, Lam J, Li P (2014) Positive filtering for continuous-time positive systems under L_1 performance. Int J Control 87(9):1906–1913

Chapter 6
Controller and Filter Syntheses for Positive Takagi–Sugeno Fuzzy Systems Under ℓ_1 Performance

In this chapter, the ℓ_1-induced controller and filter design problems for positive Takagi–Sugeno (T–S) fuzzy systems are investigated with the use of linear Lyapunov function. A novel performance characterization is first established to guarantee the asymptotic stability of the positive fuzzy system with ℓ_1-induced performance. Then, sufficient conditions are presented to design a desired controller and an iterative convex optimization approach is developed to solve the conditions. Moreover, to estimate the output of positive T–S fuzzy systems, error-bounding positive filters are constructed. Sufficient conditions expressed by linear programming problems are derived to design the desired filters. Finally, some numerical examples are presented to show the effectiveness of the derived theoretical results.

This chapter is briefly organized as follows. In Sect. 6.1, some preliminaries results of positive fuzzy systems are introduced and the performance characterization is proposed. In Sect. 6.2, the controller synthesis problem is formulated and the corresponding controller design method for positive fuzzy systems is derived. In Sect. 6.3, the filtering problem for positive fuzzy systems is formulated and the positive filter design procedure is established for positive T–S fuzzy systems. Examples are provided in Sect. 6.4 to illustrate the application of the theoretical results. The results are finally concluded in Sect. 6.5.

6.1 Performance Analysis

In this section, we first introduce notations and several results concerning positive fuzzy systems. Then, the ℓ_1-induced performance is proposed for positive fuzzy systems. Moreover, a characterization is proposed under which the positive fuzzy system is asymptotically stable and satisfies the performance in (6.4).

X. Chen, *Analysis and Synthesis of Positive Systems Under ℓ_1 and L_1 Performance*, Springer Theses, DOI 10.1007/978-981-10-2227-2_6

Consider the following fuzzy system described by the ith rule as follows:

Model Rule i: IF $\theta_1(k)$ is M_{i1} and $\theta_2(k)$ is M_{i2} and ... and $\theta_g(k)$ is M_{ig}, THEN

$$\begin{cases} x(k+1) = A_i x(k) + B_{wi} w(k), \\ y(k) = C_i x(k) + D_{wi} w(k), \end{cases} \tag{6.1}$$

where $x(k) \in \mathbb{R}^n$, $w(k) \in \mathbb{R}^m$ and $y(k) \in \mathbb{R}^q$ denote the system state, disturbance input and output, respectively. The index $i \in \{1, 2, \ldots, r\}$ gives the rule number. $\theta_1(k), \theta_2(k), \ldots, \theta_g(k)$ are the premise variables and $\theta(k) = [\theta_1(k), \theta_2(k), \ldots, \theta_g(k)]$ is the premise variables vector. M_{ie} $(i = 1, 2, \ldots, r;\ e = 1, 2, \ldots, g)$ represents the fuzzy sets. Then, we have the fuzzy system:

$$\begin{cases} x(k+1) = \sum_{i=1}^{r} h_i(\theta(k))(A_i x(k) + B_{wi} w(k)), \\ y(k) = \sum_{i=1}^{r} h_i(\theta(k))(C_i x(k) + D_{wi} w(k)), \end{cases} \tag{6.2}$$

where

$$h_i(\theta(k)) = \mu_i(\theta(k)) / \sum_{i=1}^{r} \mu_i(\theta(k)),\ \mu_i(\theta(k)) = \prod_{e=1}^{g} M_{ie}(\theta_e(k)),$$

and $M_{ie}(\theta_e(k)) \in [0, 1]$ represents the grade of membership of $\theta_e(k)$ in M_{ie}. For all k, we have

$$\sum_{i=1}^{r} h_i(\theta(k)) = 1,\ h_i(\theta(k)) \geq 0, \qquad i = 1, 2, \ldots, r.$$

Here, the following definition is given, which will be used in the sequel.

Definition 6.1 *System (6.2) is a discrete-time positive system if for all* $x(0) \succeq\succeq 0$ *and input* $w(k) \succeq\succeq 0$, *we have* $x(k) \succeq\succeq 0$ *and* $y(k) \succeq\succeq 0$ *for* $k \in \mathbb{N}$.

Next, some useful results are introduced.

Lemma 6.1 ([1]) *The discrete-time system (6.2) is positive if and only if*

$$A_i \succeq\succeq 0,\ B_{wi} \succeq\succeq 0,\ C_i \succeq\succeq 0,\ D_{wi} \succeq\succeq 0,\ i = 1, 2, \ldots, r.$$

Proposition 6.1 *System (6.2) with input* $w(k) = 0$ *is asymptotically stable if there exists a vector* $p_i \succeq\succeq 0$ *(or* $p_i \gg 0$*) satisfying*

$$p_i^T A_j - p_j^T \ll 0, \tag{6.3}$$

where $i, j = 1, 2, \ldots, r$.

Proof Consider the linear Lyapunov function $V(x(k)) = \left(\sum_{i=1}^{r} h_i(\theta(k))p_i\right)^T x(k)$ and we have

$$\begin{aligned}
\Delta V(x(k)) &= \left(\sum_{i=1}^{r} h_i(\theta(k+1))p_i\right)^T x(k+1) - \left(\sum_{i=1}^{r} h_i(\theta(k))p_i\right)^T x(k) \\
&= \sum_{i=1}^{r}\sum_{j=1}^{r} h_i(\theta(k+1))h_j(\theta(k))\left(p_i^T A_j x(k)\right) - \left(\sum_{j=1}^{r} h_j(\theta(k))p_j\right)^T x(k) \\
&= \sum_{i=1}^{r}\sum_{j=1}^{r} h_i(\theta(k+1))h_j(\theta(k)\left(p_i^T A_j - p_j^T\right)x(k)
\end{aligned}$$

From (6.3), $\Delta V(x(k)) < 0$ holds and therefore system (6.2) is asymptotically stable. □

Here, the definition of ℓ_1-induced performance is introduced. We say that a stable positive system (6.2) has ℓ_1-induced performance at the level γ if, under zero initial conditions,

$$\sup_{w \neq 0,\ w \in \ell_1(\bar{\mathbb{R}}_+^n)} \frac{\|y\|_{\ell_1}}{\|w\|_{\ell_1}} < \gamma, \tag{6.4}$$

where $\gamma > 0$ is a given scalar.

The following result is derived to serve as a characterization on the asymptotic stability of system (6.2) with the ℓ_1-induced performance in (6.4).

Proposition 6.2 *The positive fuzzy system (6.2) is asymptotically stable and satisfies* $\|y\|_{\ell_1} < \gamma\|w\|_{\ell_1}$ *if there exists a vector* $p_i \geq\geq 0$ *satisfying*

$$\mathbf{1}^T C_j + p_i^T A_j - p_j^T \ll 0, \tag{6.5}$$

$$\mathbf{1}^T D_{wj} + p_i^T B_{wj} - \gamma\mathbf{1}^T \ll 0, \tag{6.6}$$

where $i, j = 1, 2, \ldots, r$.

Proof First, we assume that $x(k) \equiv 0$. From (6.5), system (6.2) is asymptotically stable. Moreover, when $x(k) \equiv 0$, $y(k) = \sum_{i=1}^{r} h_i(\theta(k))D_{wi}w(k)$ holds and from (6.6), we have $\|y\|_{\ell_1} < \gamma\|w\|_{\ell_1}$.

Then, we assume that there exists a k such that $x(k) \neq 0$. From (6.5) and Proposition 6.1, system (6.2) is asymptotically stable.

Consider the Lyapunov function $V(x(k)) = \left(\sum_{i=1}^{r} h_i(\theta(k))p_i\right)^T x(k)$ and the following equations hold.

$$\begin{aligned}\Delta V(x(k)) &= \left(\sum_{i=1}^{r} h_i(\theta(k+1))p_i\right)^T x(k+1) - \left(\sum_{i=1}^{r} h_i(\theta(k))p_i\right)^T x(k)\\ &= \sum_{i=1}^{r}\sum_{j=1}^{r} h_i(\theta(k+1))h_j(\theta(k))\left(\left(p_i^T A_j - p_j^T\right)x(k) + p_i^T B_{wj} w(k)\right).\end{aligned}$$

Let

$$\begin{aligned}J &= \|y(k)\|_1 - \gamma\|w(k)\|_1\\ &= \mathbf{1}^T y(k) - \gamma\mathbf{1}^T w(k)\\ &= \left[\mathbf{1}^T y(k) - \gamma\mathbf{1}^T w(k) + \Delta V(k)\right] - \Delta V(k)\\ &= \sum_{i=1}^{r}\sum_{j=1}^{r} h_i(\theta(k+1))h_j(\theta(k))\left(\left(\mathbf{1}^T C_j + p_i^T A_j - p_j^T\right)x(k)\right.\\ &\quad \left. + \left(\mathbf{1}^T D_{wj} + p_i^T B_{wj} - \gamma\mathbf{1}^T\right)w(k)\right) - \Delta V(k)\\ &= \sum_{i=1}^{r}\sum_{j=1}^{r} h_i(\theta(k+1))h_j(\theta(k))\left(\left(\mathbf{1}^T C_j + p_i^T A_j - p_j^T + \varepsilon\mathbf{1}^T\right)x(k)\right.\\ &\quad \left. + \left(\mathbf{1}^T D_{wj} + p_i^T B_{wj} - \gamma\mathbf{1}^T\right)w(k)\right) - \varepsilon\mathbf{1}^T x(k) - \Delta V(k),\end{aligned}$$

where $\varepsilon > 0$ is sufficiently small such that $\mathbf{1}^T C_j + p_i^T A_j - p_j^T + \varepsilon\mathbf{1}^T \ll 0$ holds.

From (6.5) and (6.6), we have the inequality

$$J + \varepsilon\mathbf{1}^T x(k) + \Delta V(k) < 0,$$

which equals

$$\mathbf{1}^T y(k) + \varepsilon\mathbf{1}^T x(k) < \gamma\mathbf{1}^T w(k) - \Delta V(k).$$

Then, the following inequality holds

$$\sum_{k=0}^{s}\mathbf{1}^T y(k) + \varepsilon\sum_{k=0}^{s}\mathbf{1}^T x(k) < \gamma\sum_{k=0}^{s}\mathbf{1}^T w(k) - V(s+1).$$

Due to the asymptotic stability of system (6.2), when $s \to \infty$ we have

$$\sum_{k=0}^{\infty}\mathbf{1}^T y(k) + \varepsilon\sum_{k=0}^{\infty}\mathbf{1}^T x(k) \le \gamma\sum_{k=0}^{\infty}\mathbf{1}^T w(k), \tag{6.7}$$

which implies

$$\|y\|_{\ell_1} < \gamma \|w\|_{\ell_1}. \tag{6.8}$$

The whole proof is completed. □

6.2 Controller Synthesis for Positive Fuzzy Systems

In this section, the ℓ_1-induced controller synthesis problem is formulated. Based on the stability and performance conditions, a controller is designed for positive fuzzy systems. Finally, an iterative convex optimization approach is developed to solve the conditions accordingly.

Here, we deal with the stabilization problem for the following positive fuzzy systems:

$$\begin{cases} x(k+1) = \sum_{i=1}^{r} h_i(\theta(k))(A_i x(k) + B_i u(k) + B_{wi} w(k)), \\ y(k) = \sum_{i=1}^{r} h_i(\theta(k))(C_i x(k) + D_i u(k) + D_{wi} w(k)), \end{cases} \tag{6.9}$$

For the positive fuzzy system in (6.9), we construct the following parallel distributed compensation (PDC) fuzzy controller:

Control Rule i: IF $\theta_1(k)$ is M_{i1} and $\theta_2(k)$ is M_{i2} and … and $\theta_g(k)$ is M_{ig}, THEN

$$u(k) = K_i x(k).$$

The overall fuzzy controller is represented by

$$u(k) = \sum_{i=1}^{r} h_i(\theta(k)) K_i x(k). \tag{6.10}$$

Then, the closed-loop system with the PDC fuzzy controller (6.10) is given by

$$\begin{cases} x(k+1) = \sum_{i=1}^{r} \sum_{j=1}^{r} h_i(\theta(k)) h_j(\theta(k))((A_i + B_i K_j)x(k) + B_{wi} w(k)), \\ y(k) = \sum_{i=1}^{r} \sum_{j=1}^{r} h_i(\theta(k)) h_j(\theta(k))((C_i + D_i K_j)x(k) + D_{wi} w(k)), \end{cases} \tag{6.11}$$

The problem to be addressed in this chapter is described as follows.

Problem PPL1CD (Positivity-Preserving ℓ_1-induced Controller Design)
Given the fuzzy positive system (6.9), the control objective is to find a controller (6.10) such that the closed-loop system (6.11) is positive, asymptotically stable, and satisfies the ℓ_1-induced performance in (6.4) under zero initial conditions.

First, the stability characterization of the closed-loop system (6.11) is proposed as follows.

Proposition 6.3 *The closed-loop system in (6.11) with input $w(k) = 0$ is asymptotically stable if there exists a vector $p_i \geq\geq 0$ (or $p_i \gg 0$) satisfying*

$$p_i^T(A_j + B_jK_t) - p_t^T \ll 0, \tag{6.12}$$

where $i, j, t = 1, 2, \ldots, r$.

Proof Consider the Lyapunov function $V(x(k)) = \left(\sum_{i=1}^{r} h_i(\theta(k))p_i\right)^T x(k)$ and we have

$$\begin{aligned}\Delta V(x(k)) &= \left(\sum_{i=1}^{r} h_i(\theta(k+1))p_i\right)^T x(k+1) - \left(\sum_{i=1}^{r} h_i(\theta(k))p_i\right)^T x(k)\\ &= \sum_{i=1}^{r}\sum_{j=1}^{r}\sum_{t=1}^{r} h_i(\theta(k+1))h_j(\theta(k))h_t(\theta(k))\left(p_i^T(A_j + B_jK_t) - p_t^T\right)x(k).\end{aligned}$$

From (6.12), we have $\Delta V(x(k)) < 0$. Therefore, system (6.11) is asymptotically stable. □

Next, a sufficient condition for the existence of a solution to Problem PPL1CD is obtained.

Theorem 6.1 *The closed-loop system in (6.11) is positive, asymptotically stable and satisfies $\|y\|_{\ell_1} < \gamma\|w\|_{\ell_1}$ if there exist matrices K_i and vectors $p_i \gg 0$ satisfying*

$$A_i + B_iK_j \geq\geq 0, \tag{6.13}$$

$$C_i + D_iK_j \geq\geq 0, \tag{6.14}$$

$$\mathbf{1}^T(C_j + D_jK_t) + p_i^T(A_j + B_jK_t) - p_t^T \ll 0, \tag{6.15}$$

$$\mathbf{1}^T D_{wj} + p_i^T B_{wj} - \gamma\mathbf{1}^T \ll 0, \tag{6.16}$$

where $i, j, t = 1, 2, \ldots, r$.

In the following, our aim is to derive a numerically tractable means to synthesize a derived controller. It is noted that when matrix K_t is fixed, (6.15) turns out to be linear with respect to the other variables. Therefore, a natural way is to fix K_t, and solve (6.15) and (6.16) by linear programming. Thus, the following iterative algorithm can be proposed to solve the problem (see [2]).

Algorithm PPL1CD

- Step 1. Set $\kappa = 1$. Select an initial matrix sK_{i1}.
  System (6.9) with $u(k) = \sum_{i=1}^{r} h_i(\theta(k))K_{i1}x(k)$ is positive and asymptotically stable if there exist matrices $X \triangleq \text{diag}(X_1, X_2, \ldots, X_n)$ and $Y_i \triangleq [Y_{i\mu\nu}] \in \mathbb{R}^{l\times n}$ such that

$$\begin{bmatrix} -X & A_iX + B_iY_j \\ * & -X \end{bmatrix} < 0,$$

$$A_{i\mu\nu}X_\nu + \sum_{z=1}^{m} B_{i\mu z}Y_{jz\nu} \geq 0,$$

$$C_{i\alpha\nu}X_\nu + \sum_{z=1}^{m} D_{i\alpha z}Y_{jz\nu} \geq 0.$$

  Under these conditions, an initial choice of K_i can be given by $K_{i1} = Y_iX^{-1}$.
- Step 2. For fixed $K_{i\kappa}$, solve the following optimization problem for $p_{i\kappa}$ and γ_κ.
  OP1: Minimize γ_κ subject to the following constraints:

$$\begin{aligned} \mathbf{1}^T(C_j + D_jK_{t\kappa}) + p_{i\kappa}^T(A_j + B_jK_{t\kappa}) - p_{t\kappa}^T &\ll 0, \\ \mathbf{1}^TD_{wj} + p_{i\kappa}^TB_{wj} - \gamma_\kappa\mathbf{1}^T &\ll 0, \\ p_{i\kappa} &\gg 0. \end{aligned}$$

  Denote γ_κ^*, $p_{i\kappa}$ as the solution to the optimization problem. If $\left|\left(\gamma_\kappa^* - \gamma_{\kappa-1}^*\right)/\gamma_\kappa^*\right| \leq \varepsilon_1$, where ε_1 is a prescribed bound, then $K_i = K_{i\kappa}$, $p_i = p_{i\kappa}$. STOP.
- Step 3. For fixed $p_{i\kappa}$, solve the following optimization problem for $K_{i\kappa}$.
  OP2: Minimize γ_κ subject to the following constraints:

$$\begin{aligned} A_i + B_iK_{j\kappa} &\geq\geq 0, \\ C_i + D_iK_{j\kappa} &\geq\geq 0, \\ \mathbf{1}^T(C_j + D_jK_{t\kappa}) + p_{i\kappa}^T(A_j + B_jK_{t\kappa}) - p_{t\kappa}^T &\ll 0. \\ \mathbf{1}^TD_{wj} + p_{i\kappa}^TB_{wj} - \gamma_\kappa\mathbf{1}^T &\ll 0. \end{aligned}$$

  Denote γ_κ^* as the solution to the optimization problem.
- Step 4. If $\left|\left(\gamma_\kappa^* - \gamma_{\kappa-1}^*\right)/\gamma_\kappa^*\right| \leq \varepsilon_2$, where ε_2 is a prescribed tolerance, STOP; else set $\kappa = \kappa + 1$ and $K_{i\kappa} = K_{i(\kappa-1)}$, then go to Step 2.

Remark 6.1 The parameter γ can be optimized iteratively. Notice that $\gamma_{\kappa+1}^* \leq \gamma_\kappa^*$ since the corresponding parameters obtained in Step 3 will be utilized as the initial values in Step 2 to derive a smaller γ. Therefore, the convergence of the iterative process is naturally guaranteed.

6.3 Filter Synthesis for Positive Fuzzy Systems

In this section, we consider the stable fuzzy system:

$$\begin{cases} x(k+1) = \sum_{i=1}^{r} h_i(\theta(k))(A_i x(k) + B_i w(k)), \\ y(k) = \sum_{i=1}^{r} h_i(\theta(k))(C_i x(k) + D_i w(k)), \\ z(k) = \sum_{i=1}^{r} h_i(\theta(k)) L_i x(k), \end{cases} \tag{6.17}$$

where $x(k) \in \mathbb{R}^n$, $w(k) \in \mathbb{R}^l$, $y(k) \in \mathbb{R}^q$ and $z(k) \in \mathbb{R}^q$ denote the state vector, disturbance signal, measurement and the signal to be estimated, respectively. System (6.17) is positive if for all $x(0) \geq\geq 0$ and all input $w(k) \geq\geq 0$, we have $x(k) \geq\geq 0$, $y(k) \geq\geq 0$ and $z(k) \geq\geq 0$ for $k \in \mathbb{N}$.

It is noted that we cannot obtain the information of the transient output by designing conventional filters, since they only give an estimate of the output asymptotically. To design a filter which can be used to estimate the output at all times, we intend to find a lower-bounding estimate $\check{z}(k)$ and an upper-bounding one $\hat{z}(k)$. With the two estimates, the signal $z(k)$ can be encapsulated at all times. In the following, a pair of filters is proposed as follows:

Filter Rule i: IF $\theta_1(t)$ is M_{i1} and $\theta_2(t)$ is M_{i2} and ... and $\theta_g(t)$ is M_{ig}, THEN

$$\begin{cases} \check{x}(k+1) = \sum_{i=1}^{r} h_i(\theta(k))(\underline{A}_{fi}\check{x}(k) + \underline{B}_{fi} y(k)), \\ \check{z}(k) = \sum_{i=1}^{r} h_i(\theta(k))\underline{C}_{fi}\check{x}(k), \end{cases} \tag{6.18}$$

and

$$\begin{cases} \hat{x}(k+1) = \sum_{i=1}^{r} h_i(\theta(k))(\overline{A}_{fi}\hat{x}(k) + \overline{B}_{fi} y(k)), \\ \hat{z}(k) = \sum_{i=1}^{r} h_i(\theta(k))\overline{C}_{fi}\hat{x}(k), \end{cases} \tag{6.19}$$

where $\hat{x}(k) \in \mathbb{R}^n$, $\check{x}(k) \in \mathbb{R}^n$, $\hat{z}(k) \in \mathbb{R}^q$ and $\check{z}(k) \in \mathbb{R}^q$. $\underline{A}_{fi}$, $\overline{A}_{fi}$, $\underline{B}_{fi}$, $\overline{B}_{fi}$, $\underline{C}_{fi}$ and $\overline{C}_{fi}$ are filtering parameters.

First, the lower-bounding case is considered. With new variables $\check{x}_e(k) = x(k) - \check{x}(k)$, $\check{\xi}(k) = [x^T(k),\ \check{x}_e^T(k)]^T$ and $\check{e}(k) = z(k) - \check{z}(k)$, from systems (6.17) and (6.18), we have the augmented system as follows:

$$\begin{cases} \check{\xi}(k+1) = \sum_{i=1}^{r}\sum_{j=1}^{r} h_i(\theta(k))h_j(\theta(k))(\underline{A}_{\xi ij}\check{\xi}(k) + \underline{B}_{\xi ij} w(k)), \\ \check{e}(k) = \sum_{i=1}^{r} h_i(\theta(k))\underline{C}_{\xi i}\check{\xi}(k), \end{cases} \tag{6.20}$$

where

$$\underline{A}_{\xi ij} = \begin{bmatrix} A_i & 0 \\ A_i - \underline{B}_{fi}C_j - \underline{A}_{fi} & \underline{A}_{fi} \end{bmatrix},$$
$$\underline{B}_{\xi ij} = \begin{bmatrix} B_i \\ B_i - \underline{B}_{fi}D_j \end{bmatrix}, \ \underline{C}_{\xi i} = \begin{bmatrix} L_i - \underline{C}_{fi} & \underline{C}_{fi} \end{bmatrix}. \tag{6.21}$$

The filter (6.18) is designed to approximate $z(k)$ with $\check{z}(k)$. Consequently, the estimate $\check{z}(k)$ is required to be positive, which implies that the filter (6.18) is supposed to be a positive system. From Lemma 6.1, we see that $\underline{A}_{fi} \geq\geq 0$, $\underline{B}_{fi} \geq\geq 0$ and $\underline{C}_{fi} \geq\geq 0$ are needed. In the following, the fuzzy positive lower-bounding filtering (FPLF) problem is formulated.

Fuzzy Positive Lower-bounding Filtering (FPLF): *Given a stable fuzzy positive system (6.17), find a fuzzy positive filter (6.18) with* $\underline{A}_{fi} \geq\geq 0$, $\underline{B}_{fi} \geq\geq 0$ and $\underline{C}_{fi} \geq\geq 0$ *such that the filtering error system (6.20) is positive, asymptotically stable and satisfies the performance* $\|\check{e}\|_{\ell_1} < \gamma_l \|w\|_{\ell_1}$ *under zero initial conditions.*

Similarly, one may define $\hat{x}_e(k) = \hat{x}(k) - x(k)$, $\hat{\xi}(k) = [x^T(k),\ \hat{x}_e^T(k)]^T$ and $\hat{e}(k) = \hat{z}(k) - z(k)$, the filtering error system is formulated as follows:

$$\begin{cases} \hat{\xi}(k+1) = \sum_{i=1}^{r}\sum_{j=1}^{r} h_i(\theta(k))h_j(\theta(k))(\overline{A}_{\xi ij}\hat{\xi}(k) + \overline{B}_{\xi ij}w(k)), \\ \hat{e}(k) = \sum_{i=1}^{r} h_i(\theta(k))\overline{C}_{\xi i}\hat{\xi}(k), \end{cases} \tag{6.22}$$

where

$$\overline{A}_{\xi ij} = \begin{bmatrix} A_i & 0 \\ \overline{A}_{fi} + \overline{B}_{fi}C_j - A_i & \overline{A}_{fi} \end{bmatrix},$$
$$\overline{B}_{\xi ij} = \begin{bmatrix} B_i \\ \overline{B}_{fi}D_j - B_i \end{bmatrix}, \ \overline{C}_{\xi i} = \begin{bmatrix} \overline{C}_{fi} - L_i & \overline{C}_{fi} \end{bmatrix}.$$

Next, the fuzzy positive upper-bounding filtering (FPUF) problem is established.

Fuzzy Positive Upper-bounding Filtering (FPUF): *Given a stable fuzzy positive system (6.17), find a fuzzy positive filter (6.19) with* $\overline{A}_{fi} \geq\geq 0$, $\overline{B}_{fi} \geq\geq 0$ *and* $\overline{C}_{fi} \geq\geq 0$ *such that the filtering error system (6.22) is positive, asymptotically stable and satisfies the performance* $\|\hat{e}\|_{\ell_1} < \gamma_u \|w\|_{\ell_1}$ *under zero initial conditions.*

In the following, a pair of fuzzy positive error-bounding filters is designed to bound the signal $z(k)$ with the ℓ_1-induced performance. In detail, sufficient conditions are presented for the design of the lower-bounding filter. Then, parallel results are obtained for the upper-bounding case.

The stability characterization of the lower-bounding augmented system (6.20) is proposed as follows.

Lemma 6.2 *System (6.20) is asymptotically stable if there exists a vector* $p_i \geq\geq 0$ *(or* $p_i \gg 0$*) satisfying*

$$p_i^T \underline{A}_{\xi jt} - p_t^T \ll 0, \tag{6.23}$$

where $i, j, t = 1, 2, \ldots, r$.

Proof Consider the Lyapunov function $V(\check{\xi}(k)) = \left(\sum_{i=1}^{r} h_i(\theta(k))p_i\right)^T \check{\xi}(k)$ and we have

$$\begin{aligned}\Delta V(\check{\xi}(k)) &= \left(\sum_{i=1}^{r} h_i(\theta(k+1))p_i\right)^T \check{\xi}(k+1) - \left(\sum_{i=1}^{r} h_i(\theta(k))p_i\right)^T \check{\xi}(k)\\ &= \sum_{i=1}^{r}\sum_{j=1}^{r}\sum_{t=1}^{r} h_i(\theta(k+1))h_j(\theta(k))h_t(\theta(k))\left(p_i^T \underline{A}_{\xi jt} - p_t^T\right)\check{\xi}(k).\end{aligned}$$

From (6.23), we have $\Delta V(\check{\xi}(k)) < 0$. Therefore, system (6.20) is asymptotically stable. □

Next, the performance characterization result of lower-bounding filtering error system (6.20) is provided.

Theorem 6.2 *The filtering error system (6.20) is asymptotically stable and satisfies* $\|\check{e}\|_{\ell_1} < \gamma_l \|w\|_{\ell_1}$ *if there exists a vector* $p_i \geq\geq 0$ *satisfying*

$$\mathbf{1}^T \underline{C}_{\xi t} + p_i^T \underline{A}_{\xi jt} - p_t^T \ll 0, \tag{6.24}$$

$$p_i^T \underline{B}_{\xi jt} - \gamma_l \mathbf{1}^T \ll 0, \tag{6.25}$$

where $i, j, t = 1, 2, \ldots, r$.

Based on the performance characterization, the following theorem is presented to design a desired lower-bounding filter.

Theorem 6.3 *Given a stable fuzzy discrete-time positive system (6.17), a lower-bounding filter (6.18) exists such that the filtering error system (6.20) is positive, asymptotically stable and satisfies* $\|\check{e}\|_{\ell_1} < \gamma_l \|w\|_{\ell_1}$ *if there exist vectors* $p_{1i} \geq\geq 0$, $p_{2i} \geq\geq 0$ *and matrices* $\underline{M}_{A_{fij}}$, $\underline{M}_{B_{fij}}$, $\underline{C}_{fi} \geq\geq 0$ *satisfying*

$$\left[\underline{M}_{A_{fij}}\right]_{gv} \geq 0, \tag{6.26}$$

$$\left[\underline{M}_{B_{fij}}\right]_{gs} \geq 0, \tag{6.27}$$

$$L_i - \underline{C}_{fi} \geq\geq 0, \tag{6.28}$$

$$p_{2ig}^T \left[A_j\right]_{gv} - \left[\underline{M}_{B_{fij}}\right]_{r,g}\left[C_t\right]_{c,v} - \left[\underline{M}_{A_{fij}}\right]_{gv} \geq 0, \tag{6.29}$$

$$p_{2ig}^T \left[B_j\right]_{gs} - \left[\underline{M}_{B_{fij}}\right]_{r,g}\left[D_t\right]_{c,s} \geq 0, \tag{6.30}$$

$$\mathbf{1}^T(L_t - \underline{C}_{ft}) + p_{1i}^T A_j + p_{2i}^T A_j$$
$$-\sum\nolimits_{g=1}^{n} \left[\underline{M}_{B_{fij}}\right]_{r,g} C_t - \sum_{g=1}^{n} \left[\underline{M}_{A_{fij}}\right]_{r,g} - p_{1t}^T \ll 0, \tag{6.31}$$

$$\mathbf{1}^T \underline{C}_{ft} + \sum_{g=1}^{n} \left[\underline{M}_{A_{fij}}\right]_{r,g} - p_{2t}^T \ll 0, \tag{6.32}$$

$$p_{1i}^T B_j + p_{2i}^T B_j - \sum_{g=1}^{n} \left[\underline{M}_{B_{fij}}\right]_{r,g} D_t - \gamma_l \mathbf{1}^T \ll 0, \tag{6.33}$$

where $i, j, t = 1, 2, \ldots, r$; $g, v = 1, \ldots, n$; $s = 1, \ldots, m$. Moreover, a suitable set of $\underline{A}_{fi}$ and $\underline{B}_{fi}$ is given by

$$\left[\underline{A}_{fi}\right]_{gv} = p_{2jg}^{-1} \left[\underline{M}_{A_{fji}}\right]_{gv}, \quad \left[\underline{B}_{fi}\right]_{gs} = p_{2jg}^{-1} \left[\underline{M}_{B_{fji}}\right]_{gs}. \tag{6.34}$$

Proof Note that $p_{2i} \geq\geq 0$, it follows from (6.26), (6.27) and (6.34) that $\underline{A}_{fi} \geq\geq 0$ and $\underline{B}_{fi} \geq\geq 0$. Together with $\underline{C}_{fi} \geq\geq 0$, it implies that the fuzzy lower-bounding filter (6.18) is positive.

From (6.34) and $p_{2i} \geq\geq 0$, (6.29)–(6.30) become

$$\left[A_j\right]_{gv} - \left[\underline{B}_{fj}\right]_{r,g} \left[C_t\right]_{c,v} - \left[\underline{A}_{fj}\right]_{gv} \geq 0, \tag{6.35}$$
$$\left[B_j\right]_{gs} - \left[\underline{B}_{fj}\right]_{r,g} \left[D_t\right]_{c,s} \geq 0, \tag{6.36}$$

which indicates that

$$A_i - \underline{B}_{fi} C_j - \underline{A}_{fi} \geq\geq 0,$$
$$B_i - \underline{B}_{fi} D_j \geq\geq 0.$$

Together with $\underline{A}_{fi} \geq\geq 0$, $\underline{B}_{fi} \geq\geq 0$, $\underline{C}_{fi} \geq\geq 0$ and (6.28), from (6.21), it shows that the filtering error system (6.20) is positive.

Then, from (6.34), we have

$$\sum_{g=1}^{n} \left[\underline{M}_{A_{fij}}\right]_{r,g} = p_{2i}^T \underline{A}_{fj}, \quad \sum_{g=1}^{n} \left[\underline{M}_{B_{fij}}\right]_{r,g} = p_{2i}^T \underline{B}_{fj}. \tag{6.37}$$

With (6.37), inequalities (6.31)–(6.33) are equal to

$$\mathbf{1}^T(L_t - \underline{C}_{ft}) + p_{1i}^T A_j + p_{2i}^T A_j - p_{2i}^T \underline{B}_{fj} C_t - p_{2i}^T \underline{A}_{fj} - p_{1t}^T \ll 0,$$
$$\mathbf{1}^T \underline{C}_{ft} + p_{2i}^T \underline{A}_{fj} - p_{2t}^T \ll 0,$$
$$p_{1i}^T B_j + p_{2i}^T B_j - p_{2i}^T \underline{B}_{fj} D_t - \gamma_l \mathbf{1}^T \ll 0,$$

which further implies that

$$\mathbf{1}^T \left[L_t - \underline{C}_{ft} \; \underline{C}_{ft} \right] + p_i^T \begin{bmatrix} A_j & 0 \\ A_j - \underline{B}_{fj} C_t - \underline{A}_{fj} & \underline{A}_{fj} \end{bmatrix} - p_t^T \ll 0, \quad (6.38)$$

$$p_i^T \begin{bmatrix} B_j \\ B_j - \underline{B}_{fj} D_t \end{bmatrix} - \gamma_l \mathbf{1}^T \ll 0, \quad (6.39)$$

where $p_i^T = \left[p_{1i}^T \; p_{2i}^T \right]$.

Therefore, by Theorem 6.2, the whole proof is completed. □

For the upper-bounding case, the parallel result is presented as follows.

Theorem 6.4 *The filtering error system in (6.22) is asymptotically stable and satisfies $\|\hat{e}\|_{\ell_1} < \gamma_u \|w\|_{\ell_1}$ if there exists a vector $p_i \geq\geq 0$ satisfying*

$$\mathbf{1}^T \overline{C}_{\xi t} + p_i^T \overline{A}_{\xi jt} - p_t^T \ll 0, \quad (6.40)$$

$$p_i^T \overline{B}_{\xi jt} - \gamma_u \mathbf{1}^T \ll 0, \quad (6.41)$$

where $i, j, t = 1, 2, \ldots, r$.

Similarly as for the lower-bounding case, the following theorem is introduced to design the upper-bounding filter and the proof is omitted here.

Theorem 6.5 *Given a stable fuzzy discrete-time positive system (6.17), an upper-bounding filter (6.19) exists such that the closed-loop system (6.22) is positive, asymptotically stable and satisfies $\|\hat{e}\|_{\ell_1} < \gamma_u \|w\|_{\ell_1}$ if there exist vectors $p_{1i} \geq\geq 0$, $p_{2i} \geq\geq 0$ and matrices $\overline{M}_{A_{fij}}$, $\overline{M}_{B_{fij}}$, $\overline{C}_{fi} \geq\geq 0$ satisfying*

$$\left[\overline{M}_{A_{fij}} \right]_{gv} \geq 0, \quad (6.42)$$

$$\left[\overline{M}_{B_{fij}} \right]_{gs} \geq 0, \quad (6.43)$$

$$\overline{C}_{fi} - L_i \geq\geq 0, \quad (6.44)$$

$$\left[\overline{M}_{A_{fij}} \right]_{gv} + \left[\overline{M}_{B_{fij}} \right]_{r,g} \left[C_t \right]_{c,v} - p_{2ig}^T \left[A_j \right]_{gv} \geq 0, \quad (6.45)$$

$$\left[\overline{M}_{B_{fij}} \right]_{r,g} \left[D_t \right]_{c,s} - p_{2ig}^T \left[B_j \right]_{gs} \geq 0, \quad (6.46)$$

$$\begin{aligned} &\mathbf{1}^T (\overline{C}_{ft} - L_t) + p_{1i}^T A_j - p_{2i}^T A_j \\ &+ \sum_{g=1}^{n} \left[\overline{M}_{B_{fij}} \right]_{r,g} C_t + \sum_{g=1}^{n} \left[\overline{M}_{A_{fij}} \right]_{r,g} - p_{1t}^T \ll 0, \end{aligned} \quad (6.47)$$

$$\mathbf{1}^T \overline{C}_{ft} + \sum_{g=1}^{n} \left[\overline{M}_{A_{fij}} \right]_{r,g} - p_{2t}^T \ll 0, \quad (6.48)$$

$$p_{1i}^T B_j - p_{2i}^T B_j + \sum_{g=1}^{n} \left[\overline{M}_{B_{fij}} \right]_{r,g} D_t - \gamma_u \mathbf{1}^T \ll 0, \quad (6.49)$$

where $i, j, t = 1, 2, \ldots, r$; $g, v = 1, \ldots, n$; $s = 1, \ldots, m$. *Moreover, a suitable set of* $\overline{A}_{fi}$ *and* $\overline{B}_{fi}$ *is given by*

$$\left[\overline{A}_{fi}\right]_{gv} = p_{2jg}^{-1}\left[\overline{M}_{A_{fji}}\right]_{gv}, \quad \left[\overline{B}_{fi}\right]_{gs} = p_{2jg}^{-1}\left[\overline{M}_{B_{fji}}\right]_{gs}. \tag{6.50}$$

6.4 Numerical Examples

Two illustrative examples are presented in this section to demonstrate the effectiveness of the theoretical results.

Example 6.1 This example is given to illustrate the effectiveness of the proposed controller design approach in Theorem 6.1. Consider system (6.9) with the system matrices:

$$A = \begin{bmatrix} 0.5 + 0.1\sin x_1(k) & 0.6 \\ 0.6 & 0.4 \end{bmatrix}, \quad B = \begin{bmatrix} 0.1 \\ 0.2 \end{bmatrix},$$
$$B_w = \begin{bmatrix} 0.02 \\ 0.05 \end{bmatrix}, \quad C = \begin{bmatrix} 0.2 & 0.5 \end{bmatrix}, \quad D = 0.1, \quad D_w = 0.2.$$

The following fuzzy rules are employed.
Rule (1): IF $\theta(k) = \sin x_1(k)$ is 1, THEN

$$\begin{cases} x(k+1) = A_1 x(k) + B_1 u(k) + B_{w1} w(k), \\ y(k) = C_1 x(k) + D_1 u(k) + D_{w1} w(k), \end{cases}$$

Rule (2): IF $\theta(k) = \sin x_1(k)$ is -1, THEN

$$\begin{cases} x(k+1) = A_2 x(k) + B_2 u(k) + B_{w2} w(k), \\ y(k) = C_2 x(k) + D_2 u(k) + D_{w2} w(k), \end{cases}$$

where

$$A_1 = \begin{bmatrix} 0.6 & 0.6 \\ 0.6 & 0.4 \end{bmatrix}, \quad A_2 = \begin{bmatrix} 0.4 & 0.6 \\ 0.6 & 0.4 \end{bmatrix},$$
$$B_1 = B_2 = \begin{bmatrix} 0.1 \\ 0.2 \end{bmatrix}, \quad B_{w1} = B_{w2} = \begin{bmatrix} 0.02 \\ 0.05 \end{bmatrix},$$
$$C_1 = C_2 = \begin{bmatrix} 0.2 & 0.5 \end{bmatrix}, \quad D_1 = D_2 = 0.1, \quad D_{w1} = D_{w2} = 0.2.$$

The fuzzy model for the nonlinear system is as follows:

$$\begin{cases} x(k+1) = \sum_{i=1}^{2} h_i(\theta(k))(A_i x(k) + B_i u(k) + B_{wi} w(k)), \\ y(k) = \sum_{i=1}^{2} h_i(\theta(k))(C_i x(k) + D_i u(k) + D_i w(k)), \end{cases}$$

with

$$h_1(\theta(k)) = \frac{1 + \sin x_1(k)}{2}, \quad h_2(\theta(k)) = \frac{1 - \sin x_1(k)}{2}.$$

By solving the conditions in Theorem 6.1 using Algorithm PPL1CD, a feasible solution is achieved with

$$p_1 = p_2 = \begin{bmatrix} 2.1806 \; 2.1332 \end{bmatrix}^T,$$

which further yields the matrices of the state-feedback controller as

$$K_1 = \begin{bmatrix} -0.8160 \; -0.7097 \end{bmatrix}, \; K_2 = \begin{bmatrix} -0.2304 \; -0.7097 \end{bmatrix}.$$

In this example, we use the following external disturbance

$$w(k) = \begin{cases} 0.02, & 5 \le k \le 10, \\ 0, & \text{otherwise}. \end{cases} \tag{6.51}$$

The equilibrium points for the nonlinear system considered in this example are obtained as follows:

$$\begin{bmatrix} 0 \; 0 \end{bmatrix}^T, \begin{bmatrix} \frac{3\pi}{2} \; \frac{3\pi}{2} \end{bmatrix}^T, \begin{bmatrix} \frac{7\pi}{2} \; \frac{7\pi}{2} \end{bmatrix}^T, \ldots.$$

First, we study the stability of these equilibrium points. To determine the stability of the equilibrium point $x_e = [x_{e1} \; x_{e2}]^T$, we define a new variable $\hat{x} = x - x_e$ and obtain a new system $S(\hat{x})$. The system matrix of the linearized system of $S(\hat{x})$ around the origin is

$$\hat{A}_{x_e} = \begin{bmatrix} 0.5 + 0.1\sin(x_{e1}) + 0.1 x_{e1}\cos(x_{e1}) & 0.6 \\ 0.6 & 0.4 \end{bmatrix}.$$

According to Lyapunov indirect method, the stability of the equilibrium points can be determined by the eigenvalues of $\hat{A}_{x_e}$. For the origin, the eigenvalues of $\hat{A}_{x_e}$ are -0.1521 and 1.0521, which implies that the origin is unstable. For other equilibrium points, the eigenvalues of $\hat{A}_{x_e}$ are -0.2 and 1. No conclusions can be drawn from this situation using Lyapunov indirect method. In this case, the stability depends on the higher order terms in the Taylor series expansion for $\sin x_1(k)$ at x_{e1}. By analyzing the higher order terms, the equilibrium points, except the one at the origin, are all stable.

Next, the performances of the open-loop and the closed-loop system are investigated.

We choose the initial condition as follows:

$$x(0) = \begin{bmatrix} 0.01 \ 0.03 \end{bmatrix}^T .$$

With this choice of initial condition, Fig. 6.1 shows the response of open-loop system. It can be seen that the open-loop system is not globally asymptotically stable and it converges to the equilibrium point $[\frac{3\pi}{2} \ \frac{3\pi}{2}]^T$. Figure 6.2 shows the state response of the closed-loop system, from which we can see that the state of the closed-loop system converges to zero. Figure 6.3 shows the state-feedback controller $u(k)$.

To show the performance of the open-loop and closed-loop systems more clearly, the phase portraits are provided in Figs. 6.4 and 6.5. In these phase portraits, red circles denote the starting points or the equilibrium points.

From this example, we see that by introducing controller, the closed-loop system become asymptotically stable at the origin.

Example 6.2 To demonstrate the effectiveness of the filter design method in Theorem 6.3, we consider the following system (6.17) with the system matrices:

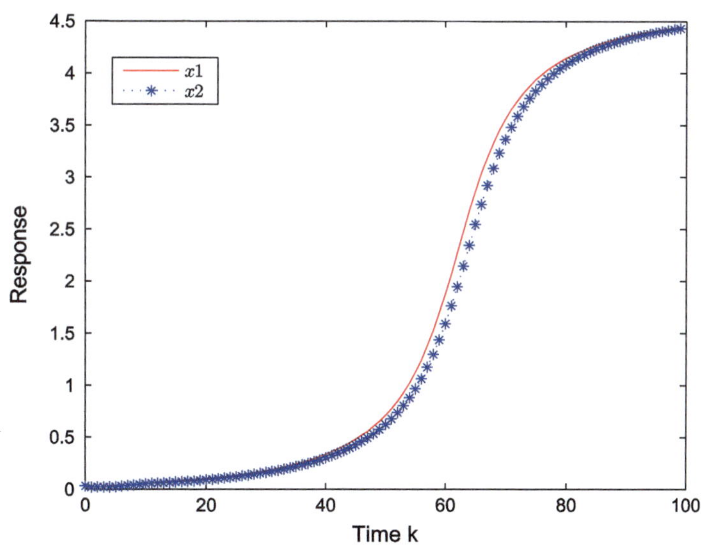

Fig. 6.1 Open-loop response under initial condition $[0.01 \ 0.03]^T$

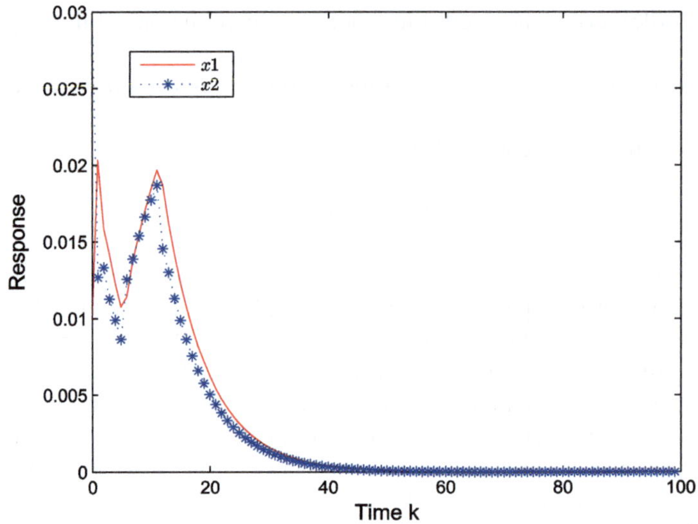

Fig. 6.2 Closed-loop response under initial condition $[0.01\ 0.03]^T$

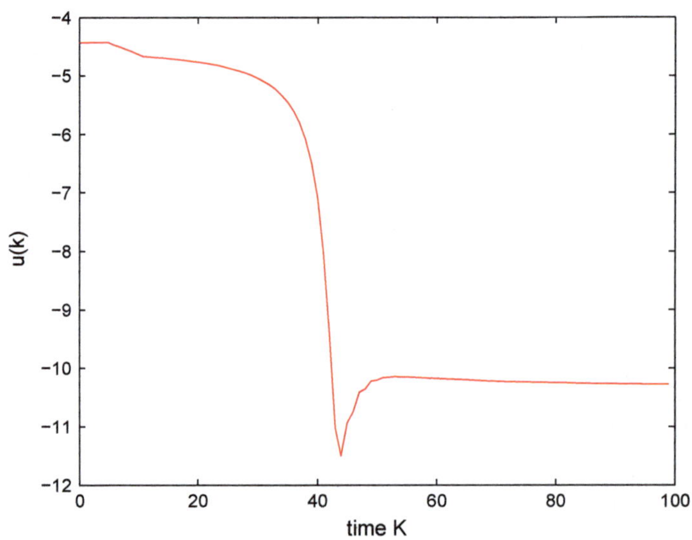

Fig. 6.3 Controller $u(k)$

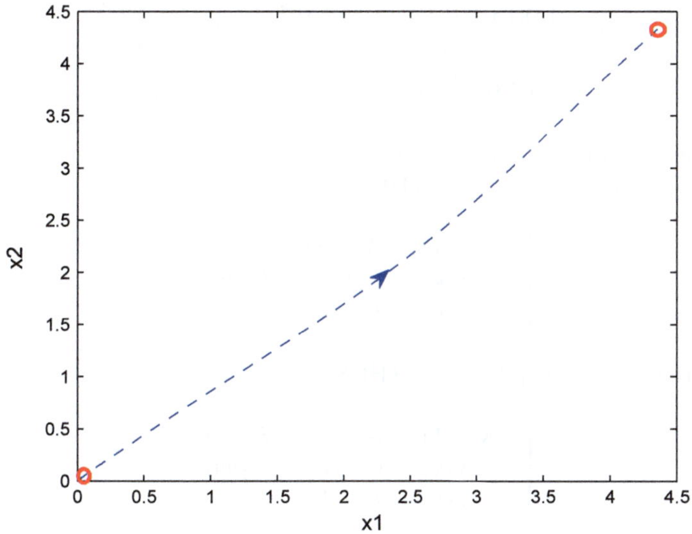

Fig. 6.4 Phase portraits of the open-loop system under initial condition $[0.01\ 0.03]^T$

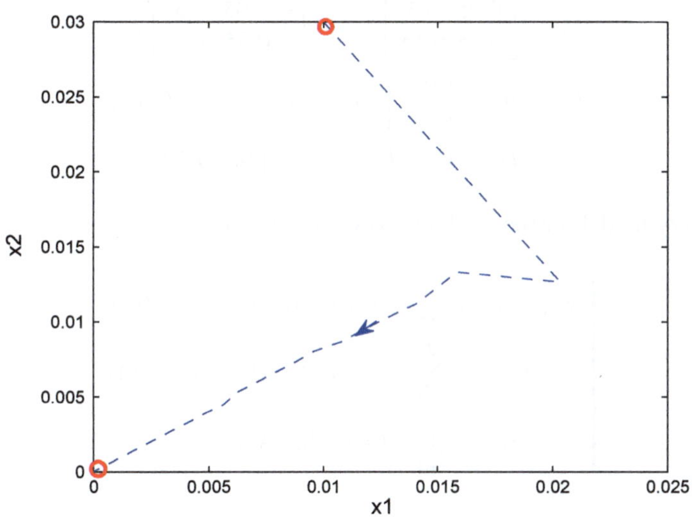

Fig. 6.5 Phase portraits of the closed-loop system under initial condition $[0.01\ 0.03]^T$

$$A = \begin{bmatrix} 0.15 + 0.1\sin x_1(k) & 0.5 \\ 0.35 & 0.15 \end{bmatrix}, \quad B = \begin{bmatrix} 0.1 \\ 0.2 \end{bmatrix},$$
$$C = \begin{bmatrix} 0.2 & 0.5 \end{bmatrix}, \quad D = 0.8, \quad L = \begin{bmatrix} 0.2 & 0.1 \end{bmatrix}.$$

Here, we use the following fuzzy rules.
Rule (1): IF $\theta(k) = \sin x_1(k)$ is 1, THEN

$$\begin{cases} x(k+1) = A_1 x(k) + B_1 w(k), \\ y(k) = C_1 x(k) + D_1 w(k), \\ z(k) = L_1 x(k), \end{cases}$$

Rule (2): IF $\theta(k) = \sin x_1(k)$ is -1, THEN

$$\begin{cases} x(k+1) = A_2 x(k) + B_2 w(k), \\ y(k) = C_2 x(k) + D_2 w(k), \\ z(k) = L_2 x(k), \end{cases}$$

where

$$A_1 = \begin{bmatrix} 0.25 & 0.5 \\ 0.35 & 0.15 \end{bmatrix}, \quad A_2 = \begin{bmatrix} 0.05 & 0.5 \\ 0.35 & 0.15 \end{bmatrix},$$
$$B_1 = B_2 = \begin{bmatrix} 0.1 \\ 0.2 \end{bmatrix}, \quad C_1 = C_2 = \begin{bmatrix} 0.2 & 0.5 \end{bmatrix},$$
$$D_1 = D_2 = 0.8, \quad L_1 = L_2 = \begin{bmatrix} 0.2 & 0.1 \end{bmatrix}.$$

The fuzzy model for the nonlinear system is as follows:

$$\begin{cases} x(k+1) = \sum_{i=1}^{2} h_i(\theta(k))(A_i x(k) + B_i w(k)), \\ y(k) = \sum_{i=1}^{2} h_i(\theta(k))(C_i x(k) + D_i w(k)), \\ z(k) = \sum_{i=1}^{2} h_i(\theta(k)) L_i x(k), \end{cases}$$

with

$$h_1(\theta(k)) = \frac{1 + \sin x_1(k))}{2}, \quad h_2(\theta(k)) = \frac{1 - \sin x_1(k)}{2}.$$

For $\gamma_l = 0.5$, by solving the conditions in Theorem 6.3 via Yalmip, a feasible solution is achieved with

$$p_{11} = p_{12} = \begin{bmatrix} 0.6257 & 0.7104 \end{bmatrix}^T,$$
$$p_{21} = p_{22} = \begin{bmatrix} 0.8924 & 0.9063 \end{bmatrix}^T,$$

which further yields the matrices of the lower-bounding filter as

$$\underline{A}_{f1} = \begin{bmatrix} 0.0035 & 0.3518 \\ 0.1942 & 0.0531 \end{bmatrix}, \ \underline{A}_{f2} = \begin{bmatrix} 0.1696 & 0.3468 \\ 0.2464 & 0.0534 \end{bmatrix},$$
$$\underline{B}_{f1} = \begin{bmatrix} 0.0260 \\ 0.0883 \end{bmatrix}, \ \underline{B}_{f2} = \begin{bmatrix} 0.0361 \\ 0.0870 \end{bmatrix}, \ \underline{C}_{f1} = \underline{C}_{f2} = \begin{bmatrix} 0.1809 & 0.0865 \end{bmatrix}.$$

Similarly, for $\gamma_u = 0.5$, by solving the conditions in Theorem 6.5, a feasible solution is achieved with

$$p_{11} = \begin{bmatrix} 0.8554 & 0.9498 \end{bmatrix}^T, \ p_{12} = \begin{bmatrix} 0.8557 & 0.9497 \end{bmatrix}^T,$$
$$p_{21} = \begin{bmatrix} 1.8449 & 1.9120 \end{bmatrix}^T, \ p_{22} = \begin{bmatrix} 2.0480 & 1.9032 \end{bmatrix}^T,$$

which further yields the matrices of the upper-bounding filter as

$$\overline{A}_{f1} = \begin{bmatrix} 0.0990 & 0.4665 \\ 0.3590 & 0.0637 \end{bmatrix}, \ \overline{A}_{f2} = \begin{bmatrix} 0.2589 & 0.4664 \\ 0.3334 & 0.0637 \end{bmatrix},$$
$$\overline{B}_{f1} = \begin{bmatrix} 0.1760 \\ 0.2983 \end{bmatrix}, \ \overline{B}_{f2} = \begin{bmatrix} 0.1762 \\ 0.2982 \end{bmatrix},$$

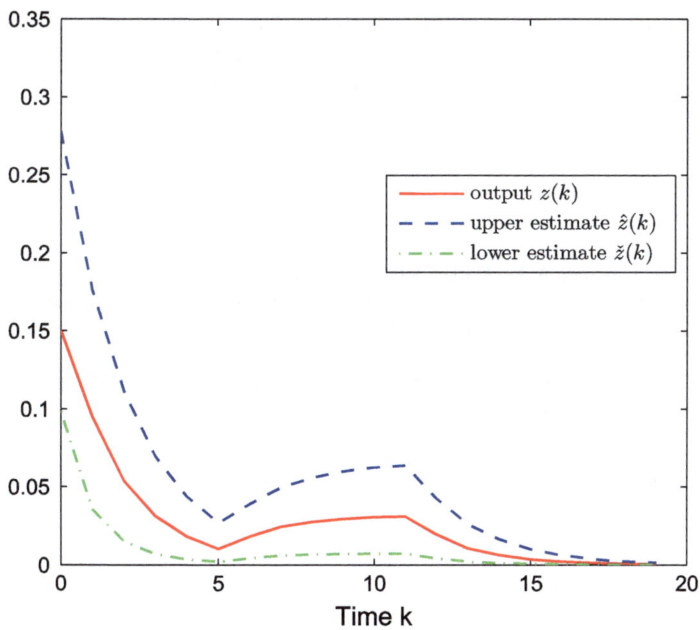

Fig. 6.6 Output $z(k)$ and its estimates. Reprinted from Ref. [3], Copyright 2015, with permission from Elsevier

$$\overline{C}_{f1} = \begin{bmatrix} 0.2617 \; 0.1578 \end{bmatrix}, \; \overline{C}_{f2} = \begin{bmatrix} 0.2620 \; 0.1576 \end{bmatrix}.$$

In this example, we use the following external disturbance

$$w(k) = \begin{cases} 0.3, & 5 \leq k \leq 10, \\ 0, & \text{otherwise}. \end{cases} \tag{6.52}$$

Figure 6.6 shows the output $z(k)$ of the original system, the lower estimate $\check{z}(k)$ and the upper estimate $\hat{z}(k)$.

6.5 Summary

In this chapter, the controller and filter design problems for positive T–S fuzzy systems have been addressed under ℓ_1 performance. Novel performance characterization of the positive fuzzy systems has been established. Based on the characterization, the controller has been designed for positive T–S fuzzy systems. In detail, sufficient conditions have been established for the existence of the desired controller. Moreover, an iterative convex optimization algorithm has been developed to solve the design conditions. In addition, the filter design problem has been investigated for positive fuzzy systems. Sufficient conditions have been developed for the existence of positive error-bounding filters. All the derived conditions are expressed by linear programming problems. Finally, numerical examples have been presented to demonstrate the effectiveness of the proposed approaches.

References

1. Benzaouia A, Hmamed A, EL Hajjaji A (2010) Stabilization of controlled positive discrete-time T-S fuzzy systems by state feedback control. Int J Adap Control Signal Process 24:1091–1106
2. Cao YY, Lam J, Sun YX et al (1998) Static output feedback stabilization: an ILMI approach. Automatica 34(12):1641–1645
3. Chen X, Lam J, Lam HK (2015) Positive filtering for positive Takagi-Sugeno fuzzy systems under ℓ_1 performance. Inf Sci 299:32–41

Chapter 7
Conclusions and Future Work

In this chapter, conclusions on the thesis are presented and some potential research issues related to the work done in this thesis are introduced.

7.1 Conclusions

This thesis is concerned with the analysis and synthesis of positive systems under ℓ_1 and L_1 performance. More specifically, two research themes have been considered.

1. For positive linear systems, the stabilization problem, state-bounding observer and filtering problems have been studied. First, the stabilization problem for positive linear systems is considered. In detail, analytical formulae to compute the exact values of ℓ_1-induced and L_1-induced norms are presented. Based on novel performance characterizations using linear programmes, necessary and sufficient conditions for the existence of desired controllers are derived. In addition, for SIMO positive systems, analytical solutions are established to show how the optimal ℓ_1-induced and L_1-induced controllers are designed. Moreover, the design of L_1-induced sparse state-feedback controller is addressed for continuous-time interval positive systems. Then, the problem of positive state-bounding observers for interval positive systems is studied under the L_1-induced performance. To estimate the state of positive systems at all times, a pair of state-bounding positive observers is designed. Necessary and sufficient conditions have been presented to design observers. Finally, the positive filtering problem is addressed for positive continuous-time systems under the L_1-induced performance. A pair of positive filters with error-bounding feature is designed to estimate the output of positive systems and the obtained results are expressed in terms of linear programming problems.
2. For positive T–S fuzzy systems, the controller synthesis and positive filtering problems have been explored under the ℓ_1-induced performance. Novel performance characterization of the positive fuzzy systems is first established. Based on the

X. Chen, *Analysis and Synthesis of Positive Systems Under ℓ_1 and L_1 Performance*, Springer Theses, DOI 10.1007/978-981-10-2227-2_7

characterization, sufficient conditions are established for the existence of state-feedback controller. An iterative convex optimization algorithm is developed to solve the design conditions. Furthermore, to estimate the output of positive T–S fuzzy systems, error-bounding positive filters are constructed. A new performance characterization is first established to guarantee the asymptotic stability of the filtering error system with the ℓ_1-induced performance. Then, sufficient conditions expressed by linear programming problems are derived to design the required filters.

7.2 Future Work

Related topics for the future research work are listed below.

1. In this thesis, the stabilization, observer, and filter design problems are investigated for positive linear systems and positive fuzzy systems under ℓ_1 and L_1 performances. In the future, it is worthwhile to deal with the analysis and synthesis problems for other types of positive systems under ℓ_1 and L_1 performances, such as positive switched systems, 2D positive systems, positive systems with delays, and Markovian jump positive systems.
2. It is worth pointing out that some constraints are introduced in the thesis to facilitate the synthesis problem. For example, we have assumed $K \leq\leq 0$ in the synthesis of controllers for interval positive systems. Moreover, in the synthesis of positive filters for positive systems, the error signal is supposed to be nonnegative. As for designing a sign-indefinite K and removing the constraint on the error signal, they remain unsolved and need further study.
3. It is interesting to consider the reachable set estimation problem for positive systems, since the bounding of reachable sets is of practical importance. For example, as stated in [1], a smaller ellipsoidal bound of reachable sets for linear systems with saturating actuators permits a larger control gain, which can result in a better performance of the system. However, this problem has not been considered for positive systems. The approach and ideas developed in the thesis can be applied to study this problem. Due to the special characteristics of positive systems, it is possible to obtain some interesting results on the reachable set estimation problem for positive systems based on linear Lyapunv functions.

Reference

1. Kim JH (2008) Improved ellipsoidal bound of reachable sets for time-delayed linear systems with disturbances. Automatica 44(11):2940–2943

MIX
Papier aus verantwortungsvollen Quellen
Paper from responsible sources
FSC® C105338

If you have any concerns about our products,
you can contact us on
ProductSafety@springernature.com

In case Publisher is established outside the EU,
the EU authorized representative is:
Springer Nature Customer Service Center GmbH
Europaplatz 3, 69115 Heidelberg, Germany

Printed by Libri Plureos GmbH
in Hamburg, Germany